PROTEOMICS

WILEY-INTERSCIENCE SERIES IN MASS SPECTROMETRY

Series Editors

Dominic M. Desiderio
Departments of Neurology and Biochemistry
University of Tennessee Health Science Center

Nico M. M. Nibbering
Vrije Universiteit Amsterdam, The Netherlands

PROTEOMICS

Introduction to Methods and Applications

Edited by

AGNIESZKA KRAJ
JERZY SILBERRING
Jagiellonian University

WILEY

A JOHN WILEY & SONS, INC., PUBLICATION

Library of Congress Cataloging-in-Publication Data:

Proteomika. English.
    Proteomics : introduction to methods and applications / [edited by]
Agnieszka Kraj, Jerzy Silberring.
        p. ; cm. – (Wiley-Interscience series on mass spectrometry)
    "Originally published in Polish, the English version will include new
contributors from around the world, updating the book with chapters on
microdevices and protein-protein interactions"–Data view.
    Includes bibliographical references and index.
    ISBN 978-0-470-05535-9 (cloth)
    1. Proteomics–Textbooks.   I. Kraj, Agnieszka.   II. Silberring, Jerzy, 1949–
III. Title.   IV. Series.
    [DNLM: 1. Proteomics–methods.   2. Proteome–analysis.   QU 58.5 P96775
2008a]
    QP551.P7568513 2008
    572'.6–dc22

                                                          2008018993

To our children: Magda, Ela, Kuba, and Marcin

CONTENTS

CONTRIBUTORS

Hesham Ali   College of Information Science and Technology, 1110 South 67th Street, Omaha, NE 68182-0116   hali@mail.unomaha.edu

Anna Bierczynska-Krzysik   Department of Neurobiochemistry, Faculty of Chemistry, Jagiellonian University, Ingardena 3, 30-060 Krakow, Poland   bierczyn@chemia.uj.edu.pl

Anna Bodzon-Kulakowska   Department of Neurobiochemistry, Faculty of Chemistry, Jagiellonian University, Ingardena 3, 30-060 Krakow, Poland   bodzon@chemia.uj.edu.pl

Pawel Ciborowski   Mass Spectrometry and Proteomics Core Facility, 985800 University of Nebraska Medical Center, Omaha, NE 68198-5800   pciborowski@unmc.edu

Anna Drabik   Department of Neurobiochemistry, Faculty of Chemistry, Jagiellonian University, Ingardena 3, 30-060 Krakow, Poland; Department of Medical Biochemistry, Medical College, Jagiellonian University, Kopernika 7, 31-034 Krakow, Poland   adrabik@chemia.uj.edu.pl

Tomasz Dylag   Department of Neurobiochemistry, Faculty of Chemistry, Jagiellonian University, Ingardena 3, 30-060 Krakow, Poland; Current address: European Commission, Office CDMA-1/022, B-1049 Brussels, Belgium   dylag@chemia.uj.edu.pl

Ibeth Guevara-Lora   Department of Analytical Biochemistry, Faculty of Biochemistry, Biophysics and Biotechnology, Jagiellonian University, Gronostajowa 7, 30-387 Krakow, Poland   ibeth@mol.uj.edu.pl

Ulf Hellman   Ludwig Institute for Cancer Research, Box 595, SE-751 24 Uppsala, Sweden   ulf.hellman@licr.uu.se

Justyna Jarzebinska Department of Neurobiochemistry, Faculty of Chemistry, Jagiellonian University, Ingardena 3, 30-060 Krakow, Poland jarzebinka@gazeta.pl

Simone König Integrated Functional Genomics, University of Muenster, Roentgenstrasse 21, 48149 Muenster, Germany   koenigs@uni-muenster.de

Andrzej Kozik Department of Analytical Biochemistry, Faculty of Biochemistry, Biophysics and Biotechnology, Jagiellonian University, Gronostajowa 7, 30-387 Krakow, Poland   akozik@mol.uj.edu.pl

Agnieszka Kraj Department of Neurobiochemistry, Faculty of Chemistry, Jagiellonian University, Ingardena 3, 30-060 Krakow, Poland kraj@chemia.uj.edu.pl

Piotr Laidler Department of Medical Biochemistry, Medical College, Jagiellonian University, Kopernika 7, 31-034 Krakow, Poland   mblaidle@cyf-kr.edu.pl

Gert Lubec Department of Pediatrics, Medical University of Vienna, Währinger Gürtel 18, A1090 Vienna, Austria   lubec@telering.at

Marcus Macht Bruker Daltonik GmbH, Fahrenheitstrasse 4, 28359 Bremen, Germany   marcus.macht@bdal.de

Adam Moszczynski Department of Neurobiochemistry, Faculty of Chemistry, Jagiellonian University, Ingardena 3, 30-060 Krakow, Poland moszczyn@chemia.uj.edu.pl

Marek Noga Department of Neurobiochemistry, Faculty of Chemistry, Jagiellonian University, Ingardena 3, 30-060 Krakow, Poland noga@chemia.uj.edu.pl

Hana Raoof Department of Neurobiochemistry, Faculty of Chemistry, Jagiellonian University, Ingardena 3, 30-060 Krakow, Poland   hraoof@chemia.uj.edu.pl

Małgorzata Rzychon Institute for Reference Materials and Measurements (IRMM), Directorate General–Joint Research Centre (DG-JRC), European Commission (EC), Retieseweg 111, 2440 Geel, Belgium; Department of Analytical Biochemistry, Faculty of Biotechnology, Jagiellonian University, Gronostajowa 7, 30-387 Krakow, Poland   malgosiarz@poczta.onet.pl

Jerzy Silberring Department of Neurobiochemistry and Regional Laboratory of Physicochemical Analyses and Structural Research, Jagiellonian University, Ingardena 3, 30-060 Krakow, Poland; Centre for Polymer and Carbon Materials, Polish Academy of Sciences, Sklodowskiej-Curie 34, 41-819 Zabrze, Poland Silber@chemia.uj.edu.pl

Filip Sucharski Department of Neurobiochemistry, Faculty of Chemistry, Jagiellonian University, Ingardena 3, 30-060 Krakow, Poland fillip.sucharski@gmail.com

Piotr Suder   Department of Neurobiochemistry and Regional Laboratory of Physicochemical Analyses and Structural Research, Faculty of Chemistry, Jagiellonian University, Ingardena 3, 30-060 Krakow, Poland   suder@chemia.uj.edu.pl

Małgorzata Iwona Szynkowska   Institute of General and Ecological Chemistry, Technical University of Lodz, Zeromskiego 116, 90-924 Lodz, Poland   miszynk@p.lodz.pl

PREFACE

The preceding decade brought spectacular developments in human genome sequencing. *Proteomics* (from "*prote*in complement of the gen*ome*") is aimed at the analysis of the entire protein content of a cell, tissue, or organism, not a simple identification of cell components. Proteomics was established in 1985 as a set of methodologies to resolve protein sequences on a large scale. Present challenges include proteins function(s) and interactions of protein network(s).

This rapidly expanding strategy integrates several disciplines, including fast and ultrasensitive methods for protein separation, bioinformatics, materials engineering, crystallography, and spectroscopy. The wide range of uses emphasizes the fact that proteomics is not just mass spectrometry, as is sometimes thought, but a link between genes and proteins and their functions, and a disease. The long-term goal is to find markers of various diseases in the hope to design new, more effective, personalized therapies. The rationale behind the concept is that pathogenic mechanisms cause changes in protein content, and the discovery of such differences between healthy and disease states may lead to a better understanding of the molecular mechanisms that occur in the cells. Thus, the main beneficiaries would be the life sciences, the pharmaceutical industry, and most of all, us — *Homo sapiens*.

This monograph is dedicated primarily to undergraduates and graduates (chemistry, biochemistry, pharmacy, medicine) but also to those interested in this continuously developing strategy. Some chapters are more detailed, as they describe new, important directions. There are also descriptions of several laboratory practices that can be used in any student's laboratory to improve practical skills in this methodology.

We do hope that this book will be a first step toward understanding the background and trends in modern proteomics. We would be grateful for feedback, and all comments will be welcome: kraj@chemia.uj.edu.pl; silber@chemia.uj.edu.pl.

Acknowledgments

Preparation of this book has been supported partially by International Centre for Genetic Engineering and Biotechnology grant CRP/POL05-02.

AGNIESZKA KRAJ
JERZY SILBERRING

1

INTRODUCTION TO PROTEOMICS AND STRATEGY FOR PROTEIN IDENTIFICATION

JERZY SILBERRING

WHY PROTEOME?

Proteomics differs, in many aspects, from other traditional methods for the isolation and identification of proteins. Using classical methods, one protein is usually isolated and an effort is made to determine its sequence, structure, and function. In contrast, proteomics deals with the simultaneous (global) analysis of all proteins, using several analytical techniques. An example of a typical working scheme is shown in Figure 1.

These two approaches also differ in respect to the instrumentation used and to the scale of analysis. It should, however, be emphasized that "classical" methods for protein characterization are still in common use, and the information on isolated single molecules obtained is much more accurate and specific. Identification procedures used in proteomics are not free of artifacts and is still concerned with the problem of inaccurate bioinformatics tools and incomplete protein databases. In many cases, even less accurate but rapid identification of possible differences in protein patterns may be the basis for further, more detailed experiments. An outstanding success of proteomics is the recent discovery that HIV-infected humans who are resistant to AIDS development have a significantly higher level of defensins, cyclic peptides produced by the immune system.

The flow of information within a cell depends on many factors, such as the protein biosynthesis pathways, whose general scheme is presented in Figure 2. Analysis of the

Proteomics: Introduction to Methods and Applications, Edited by Agnieszka Kraj and Jerzy Silberring
Copyright © 2008 John Wiley & Sons, Inc.

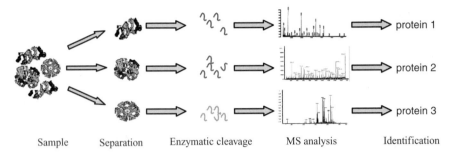

Sample	Separation	Enzymatic cleavage	MS analysis	Identification

FIGURE 1   General strategy for protein identification. (*See insert for color representation of figure.*)

genome appeared to be insufficient for identification of possible causes for the occurrence of various changes in cell content. This can be visualized as shown in Figure 3.

There is no straight-line relationship between gene expression and the most important parameters characterizing proteins. For example, gene expression in the central nervous system is relatively high but the level of particular proteins is very low. Moreover, only 10% of all genes code for specific proteins. Information in the genes does not allow us to determine posttranslational modifications. A similar problem is related to protein function. Therefore, there is an urgent need for complementary investigations using the proteomic approach. This strategy makes possible additional verification of the data obtained and leads to better control of pharmacological intervention in pathophysiology.

Moreover, information provided by genes does not include the presence of low-molecular-mass components. Most often, such substances, including nitric oxide,

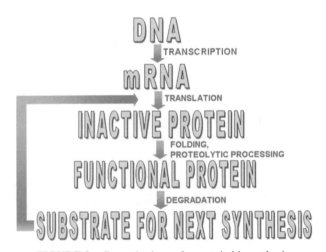

FIGURE 2   General scheme for protein biosynthesis.

FIGURE 3   Relationships between gene expression and proteomics.

biogenic amines, prostaglandins, steroid hormones, and carbon oxide, fulfill several important functions in our organisms.

CURRENT CHALLENGES

In the postgenomic era in which almost the entire DNA sequence has been determined, the next challenge is to specify the function of each gene and to determine the role of each protein encoded by the genes. In this area, termed *functional proteomics*, identification of the primary structure (amino acid sequence) is not sufficient but is absolutely necessary in assigning the role(s) of particular proteins. It often happens that the same protein may have several functions in the same organism. For example, hemoglobin is an oxygen transporter but also transfers CO_2 to the lungs. Moreover, this protein releases shorter fragments (peptides) during proteolytic cleavage. These peptides have opioidlike properties (e.g., hemorphins) or antibacterial activity (e.g., hemocidins). It is worth noting that transgenic mice, which lack myoglobin, the protein responsible for oxygen storage in muscles, do not show any symptoms and appear to feel quite happy. It is anticipated that the function of myoglobin has been replaced by another compensatory mechanism(s). Therefore, the commonly accepted dogma of one gene–one protein–one function is no longer true. Another example is that of the proopiomelanocortin (POMC) precursor, which releases adrenocorticotrophic hormone, melanotrophic hormone, and endorphins. Each of these substances shows a distinct function in the organism, but all of them are released from the same POMC precursor.

Based on information from gene sequences, we can roughly estimate that the human body contains some 60,000 to 70,000 proteins. Taking into consideration posttranslational modifications (over 100 identified!) and alternative splicing, we may end up with an approximate total of 700,000 proteins. One cell contains an average of approximately 10,000 proteins, with their quantity and proportions subject to changes in space and time. Isolation and identification of a particular protein, with its search for all modifications, its three-dimensional structure, and the function of such a huge number of molecules is a real challenge for today's scientists. It is worth noting that concentrations of endogenous compounds are often extremely low, as shown in Table 1.

We reiterate that the protein content in cells is not equal and that it changes in time and space (i.e., undergoes dynamic changes). This fact causes additional methodological problems as well as problems concerning interpretation of the results.

TABLE 1   Approximate Amounts of Substances in Living Organisms

Compound	Concentration Range
Hormones	nanomolar (10^{-9} M)
Neurotransmitters	picomolar (10^{-12} M)
Neuropeptides	femtomolar (10^{-15} M)

THE BASICS OF PROTEOMICS

As mentioned earlier, the number of proteins present in living organisms, the diversity of cells and tissues, and the dynamic changes in protein levels not only in time but also during pathological processes further complicate the methodology. Moreover, various strategies for solving these problems are not unified; that is, each laboratory may obtain a distinct set of data from the same biological material. It is also worth noting that the number of methodologies used in proteomics lends itself to the production of false-positive results. This is due to the fact that the vast amount of raw data, the output of the analytical instruments, *always* displays a "positive" result. To limit artifacts, it is advisable to focus the range of investigations and models used. For example, one can analyze proteome in prostate cancer, proteome in Down's syndrome, or peptidome in neuropeptide research. The separate aspect is *metabolomics*, a proteomics variant that deals with low-molecular-mass compounds and their metabolites. Similarly, in drug-dependence studies we can specify, for example, a morphinome or a "cocainome" because morphine and cocaine have distinct mechanisms of action on the central nervous system, inducing various changes in an organism.

Proteomics is usually associated with mass spectrometry (MS). Indeed, MS techniques are one of the most important elements of the entire strategy aimed toward the identification of proteins. But in this case, identification of a particular protein is based only on a determination of its incomplete amino acid sequence (partial sequence or peptide map). The more detailed analysis must include a full amino acid sequence, possible mutations, posttranslational modifications, tertiary and quaternary structures, interactions with other molecules, and the protein's role in the organism. It can easily be seen that from such a point of view, proteomics is not and cannot be associated with mass spectrometry only but must be linked to a vast number of techniques, including molecular biology and genetics (e.g., transgenic animals, antisense probes), crystallography, pharmacology, material engineering, and others. Additionally, analytical problems are even more complex, due to the fact that we are not able to judge which concentration of a given protein is still "physiological" or how large the differences between the physiology and pathological states should be before considering such a molecule to be of value as a marker. Sometimes, a 20% elevated protein level suggests abnormal processes, and sometimes, protein concentration must exceed two to three times its basal value before indicating pathology.

STRATEGY FOR PROTEOMIC ANALYSIS

Despite the rapid development of modern analytical techniques, simultaneous and accurate identification of the huge number of known proteins is still very difficult, if not impossible. Therefore, as mentioned earlier, laboratories try to focus such analysis on particular problems, such as cancer, drug dependence, or neurodegenerative disease. In all aspects, such investigations demand ultrasensitive and precise analytical methodologies and a solid knowledge, often at the interdisciplinary level. Because of the limited amount of precious biological material and a need for its concentration, the entire analytical process is commonly performed in a single drop (i.e., less than 5 μL).

A general strategy for protein analysis that is utilized in proteomics is shown in Figure 4. The work flow consists of several phases, each of which may be critical to the overall success of the analysis. The major challenge during the identification of endogenous compounds is the work at the very edge of the sensitivity limit, which, again, is associated with the availability of biological material. One of the crucial points is nonspecific adsorption of samples on tube walls, columns, pipette tips, and so on. All important aspects of the methodologies used in proteome research are covered in the following chapters.

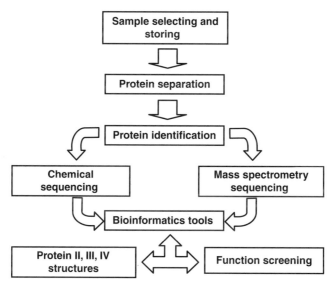

FIGURE 4   Schematic of the work flow in proteomic analysis.

2

SAMPLE PREPARATION

Agnieszka Kraj and Jerzy Silberring

The quality and reliability of the information gained from proteomic analysis both depend critically on the quality of the sample preparation step. Therefore, the proteomic approach to dealing with accurate identification of proteins, their interactions, and differences between healthy and disease states demands very precise, unambiguous, and reliable analytical methodologies. We must emphasize here that sample preparation determines all further procedures and protocols, and should be planned thoroughly. In this chapter we focus on major aspects of the preparation of biological material. We must emphasize here that despite rapid progress, there is still no unified proteomic technology platform or strategy, nor any established performance criteria.

BIOLOGICAL MATERIAL

The analytic process starts with the withdrawal or extraction of biological material. The general idea during this and later steps is to minimize the number of procedures and simplify them. All procedures should be performed at 4°C, and the material obtained should be kept at −80°C. As repetitive freezing and thawing should be avoided, the sample should be divided into smaller portions. Highest-purity chemicals should be used. Samples will be concentrated during the following purification phases, and any (even small) impurities will be concentrated together with the sample.

PRECAUTIONS

Human material should be treated with special care from the point of view of possible infections. Such samples should be taken from verified sources (e.g., laboratories,

Proteomics: Introduction to Methods and Applications, Edited by Agnieszka Kraj and Jerzy Silberring
Copyright © 2008 John Wiley & Sons, Inc.

clinics) with brief information on their history and disease, procedure of extraction, anticoagulants, inhibitory cocktails, and so on. Vaccination against hepatitis is
strongly recommended for the laboratory staff. Disposable gloves are a requirement,
as are all necessary permits for work with biological material, ethical permits, standardized and approved routines for disposal of biological or infectious samples, and
other factors.

> There is no single universal procedure for sample preparation. Each sample has its
> own specificity, and the entire protocol must be adapted to the unique conditions.

SEPARATION STRATEGIES

Separate protocols are needed for tissues, body fluids, cell cultures, and other material.
For example, brain tissue differs significantly from other tissues, as it contains a high
percentage of lipids, which strongly affects chromatographic separations. Samples
should be prepared based on the final goal, which might be simple sequencing,
retention of biological activity, or possible interactions.

> It is absolutely necessary to determine protein content before a sample is processed
> further.

The major steps in sample preparation concern:

• Isolation and withdrawal
• Extraction
• Separation
• Desalting and concentration

The general scheme is shown in Figure 1.

Isolation and withdrawal should be performed accurately, without contaminating
neighboring cells or blood. Bodily fluids (e.g., blood, cerebrospinal fluid) are of
particular interest, as they serve as a useful diagnostic source and often reflect early
changes in the pathophysiology of various diseases. Cerebrospinal fluid that is even
slightly contaminated with blood should be rejected.

The extraction procedure depends on the nature of the material. Tissues are usually
homogenized in an ice-cold buffer with the addition of an inhibitory cocktail to avoid
unwanted proteolytic degradation. Peptides are generally extracted from tissues after
boiling in 1 M acetic acid, followed by tissue homogenization. An efficient alternative
is microwave treatment of the animal, a procedure that inactivates proteolysis.

The simplest method, often used for large volumes of samples (urine), is the
salting-out technique, using ammonium sulfate and a stepwise gradient. Samples are

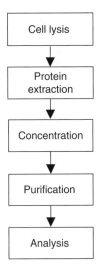

FIGURE 1   General protocol for sample preparation.

effectively concentrated and preseparated, but a challenge may be the need for a desalting procedure prior to further analytical steps.

The separation step often consists of several procedures, depending on the complexity of a sample. The proteomics approach utilizes two-dimensional electrophoresis as a most commonly used methodology, with good separation efficiency and high resolution. However, reproducibility and time of analysis are still questionable. Alternatively, two-dimensional liquid chromatograph/mass spectrometry (2-D LC/MS) is used with a combination of a cation exchanger and reversed-phase columns. Other protocols involve isoelectric focusing in solution, followed by chromatography. Many combinations of methodologies have recently been published, and they are adjusted to individual sample types, availability of techniques, and so on.

Desalting and concentration of the sample in its final form is usually performed in one step utilizing reversed-phase minicolumns and solid - phase extraction (SPE). Such devices (Figure 2) are capable of concentrating sample to a few microliters, which improves the sensitivity of measurements significantly. Such minicolumns are available commercially and can also be prepared in the laboratory using pipette tips and a bulk packing material. The format of SPE columns and cartridges is suitable for work with biological samples (limited volume). There are micro-SPE devices much as 96-well column plates or microcolumns that allow handling microliters of fluid (2 μL).

The most efficient method utilizes affinity chromatography with the sample specifically enriched with the desired component, depending on the antibody (or other binding molecule) used. This method can also serve in an opposite way to deplete biological material with unwanted components, such as the most abundant proteins in plasma. There are, however, reports that describe the disadvantages of such methods, as many other important components are also removed.

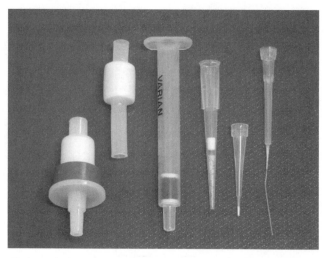

FIGURE 2   Various devices for solid-phase extraction. At the far right is a homemade microcolumn prepared from a pipette tip and filled with reversed-phase material.

SOLID-PHASE EXTRACTION

As mentioned above, solid-phase extraction constitutes an important preparation step prior to sample injection into the LC system. The main function of SPE cartridges is to separate the analyte from unwanted matrix material. This might be salt, detergent, or other components present in the sample. The characteristic feature of the SPE method is the size of the beads (ca. 40 μm or more), which is larger than the bead size used in analytical columns.

There are two types of SPE modes, online and off-line, as part of the analytical system. In the *online* approach, the enrichment and stripping modes can be used. In the enrichment mode the analyte is retained on a trap column (in most cases, reversed-phase material) and enriched before injection to the analytical column. Contaminating matrix is eluted to waste. In the stripping mode, the matrix is adsorbed on the SPE column, and the analyte is eluted and separated on the analytical column. All operations can be programmed with the help of automated switching valve systems.

A variety of stationary phases is applied for SPE, as reported above. Reversed-phase is the most common and is suitable for less polar compounds (stationary phase is hydrophobic, polar mobile phases), but also normal phase, ion-exchange, and adsorption packings are available (see also Chapters 3.1 and 3.2). Mixed-mode stationary phases, e.g., combining ion-exchange properties and reversed-phase chromatography, are also used.

In addition to the most commonly used stationary phases, there are the less popular but very promising restricted access material (RAM) and molecularly imprinted polymers (MIPs). The RAM system combines an outer surface acting as a size-exclusion filter, and an internal surface, which adsorbs low-molecular-mass molecules. Due to their exclusion from the pores, proteins have no access to the internal surface of the cartridge and are eluted. This makes it possible to separate proteinaceous contaminants from other small-molecular-mass compounds.

MIP is a polymer-based affinity type of stationary phase that functions as a structure-specific method. MIP is constructed to mimic analyte on a desired polymeric template, and eventually, stronger binding to the stationary phase is observed. As it is based on polymers, MIP packing is also stable in a wide range of pH, organic solvents, and temperature. A real challenge is extraction of highly hydrophobic or "extreme" proteins from plasma membrane and those having high or low isoelectric point values. Several recent publications describe the use of various detergents and modification of the two-dimensional PAGE procedure, adjusted to such requirements. Otherwise, for efficient extraction, CHAPS (3-[(3-cholamidopropyl)dimethylammonio]-1-propanesulfonate), Triton X-100, Nonidet, or SDS (sodium dodecyl sulfate) is used as a supplement. Those reagents usually obscure mass spectra and should be removed before final analysis.

MINIATURIZATION

The challenge in proteomics and more generally, in the discovery of endogenous molecules centers around the size of the biological sample. The material should be concentrated after purification, due to the fact that LC systems and mass spectrometers are mass detectors (i.e., the higher the concentration of a substance, the better the response of the detector). This results in miniaturization of both LC systems and ion sources (the part of the mass spectrometer responsible for efficient sample introduction into the analyzer (for a detailed description of mass spectrometry in the proteomics area, see Chapters 7.1 to 7.6). Currently, nano-LC systems, which maintain flow rates starting as long as 20 nL/min, optimized for 50 to 150-μm-inside diameter (i.d.) columns connected to a nanoelectrospray source, are the most robust analytical platform in the proteomics field.

SAMPLE LOSS

A major obstacle leading to negative results is loss of sample during its transfer between various preparation steps. A small amount of biological material can be adsorbed almost anywhere. Pipette tips, assay tubes, solid-phase cartridges, transfer tubings, and chromatographic columns are only a few examples of locations where sample can be adsorbed irreversibly. For example, a peptide map at a low femtomolar level, prepared before a weekend and kept at −80°C, would usually be undetectable after a few days of storage. The recovery yield can be improved to some extent by using siliconized tubes and pipette tips and reducing the number of preparation steps.

Acknowledgments

This work was supported by International Centre for Genetic Engineering and Biotechnology (ICGEB) grant CRP/POL05-02.

RECOMMENDED READING

Bodzon-Kulakowska A., Bierczynska-Krzysik A., Dylag T., et al. *J. Chromatogr. B Anal. Technol. Biomed. Life Sci.* 849 (2007) 1–31.

Boos K.S., Fleischer K.T. Multidimensional on-line solid-phase extraction (SPE) using restricted access materials (RAM) in combination with molecular imprinted polymers (MIP). *Fresenius J. Anal. Chem.* 371 (2001) 16–20.

Canas B., Pineiro C., Calvo E., Lopez-Ferrer D., Gallardo J.M. Trends in sample preparation for classical and second generation proteomics. *J. Chromatogr. A* 1153 (2007) 235–258.

Paulson L., Persson R., Karlsson G., Proteomics and peptidomics in neuroscience: experience of capabilities and limitations in a neurochemical laboratory. *J. Mass Spectrom.* 40 (2005) 202–213.

Speicher K.D., Kolbas O., Harper S., Speicher D.W. Systematic analysis of peptide recoveries from in-gel digestions for protein identifications in proteome studies. *J. Biomol. Technol.* 11 (2000) 74–86.

3

CHROMATOGRAPHIC SEPARATION METHODS

3.1

LIQUID CHROMATOGRAPHY LINKED TO MASS SPECTROMETRY

PIOTR SUDER

Liquid chromatography is widely used in proteomics as a preseparation technique followed by other purification methods, or as a principal methodology for the separation of complex mixtures (e.g., multidimensional chromatography). The technique used most extensively is high-performance liquid chromatography (HPLC), but traditional low-pressure techniques are still used to prepare samples for further analysis. Basic advantages of HPLC over low-pressure methods are:

1. Shorter analysis time
2. Higher sensitivity
3. Higher resolution
4. Lower sample consumption

At present, HPLC systems are usually connected to mass spectrometers, which serve as specific and sensitive detectors in many areas of research, diagnostics, and technological processes. Such a connection provides simultaneous detection and identification of compounds that cannot be collected otherwise. Mass spectrometers also provide detailed structural data.

OFF-LINE CONNECTIONS

From the point of view of the solvents and flow rates used, HPLC is, in most cases, compatible with online connections to electrospray ionization (ESI) sources. Off-line

Proteomics: Introduction to Methods and Applications, Edited by Agnieszka Kraj and Jerzy Silberring
Copyright © 2008 John Wiley & Sons, Inc.

analyses are rather rare in modern laboratories, as mass spectrometers and HPLC columns with narrow diameters are more accessible. However, in some cases, an online analysis is difficult to perform, a good example being ion-exchange chromatography. This type of column needs a salt or pH gradient for separation and following elution of the components. Involatile solvents containing salts or rapid pH changes during analysis are unacceptable for ESI sources because of the ion suppression effect or instrument clogging (e.g., phosphate buffers). The most common technique to use to remove interfering salts is desalting on microcolumns. Several commercial manufacturers offer diverse inexpensive devices filled with various stationary phases. The procedure leads to the removal of salts and detergents and also preconcentrates sample before injection. The latter feature is very important, as all mass spectrometers are mass-sensitive detectors (i.e., the higher the sample concentration in a given volume, the higher the signal that can be detected). Therefore, minimizing the sample volume is a key step in sample preparation prior to final MS analysis. This effect can also be achieved using capillary columns.

The matrix-assisted laser desorption/ionization (MALDI) ion source is an alternative to electrospray when off-line separation is necessary. LC-MALDI utilizes spotters, thus depositing samples directly on a MALDI target. The off-line connections are easier to achieve, as they do not demand a thorough practical knowledge of the entire LC-MS system. Often, samples can be preseparated in the laboratory and the fractions submitted to another facility that has expertise in mass spectrometry. Linking several instruments together is always a compromise between the optimal performance of each.

ONLINE CONNECTIONS

Online connections between separation techniques and mass spectrometry gain an increase in information received from the analysis, but most of all, they benefit from reduced sample loss and full automation. The basic advantages of MS used as a detector in HPLC are:

1. Very high sensitivity (1×10^{-11} to 1×10^{-14} M)
2. Very high resolution toward eluted compounds
3. Very high structure specificity
4. Sequencing capability
5. Higher throughput

The method of choice, which is used for on-line HPLC-MS connections, is electrospray ionization. ESI works under atmospheric pressure and is able to ionize effectively substances and broad mass ranges dissolved in liquids. These advantages make ESI sources ideal interfaces between liquid chromatographs and mass spectrometers.

Another ionization method, working at atmospheric pressure and suitable mainly for small molecules, is atmospheric pressure chemical ionization (APCI). Figure 1

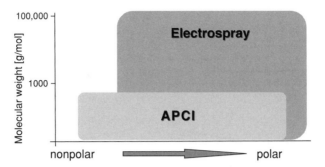

FIGURE 1   Use of ESI and APCI ionization techniques in the context of molecular mass and polarity of molecules analyzed.

presents the measuring ranges for ESI and APCI, including the hydrophobicity of compounds.

For successful use of ESI, the following rules are critical:

1. The eluent flow rate on a chromatographic column must be adjusted.
2. The eluent must be tolerated by ESI.
3. Dead volumes in the system must be eliminated.
4. Optimization of ESI parameters must be performed.

DATA INTERPRETATION

As noted earlier, mass spectrometers provide unique structural identification information not available with any other detector. Typical analytical features of a conventional ultraviolet–visible light (UV/VIS) detector compared to a mass spectrometer are shown in Figure 2. Mass spectrometers can analyze coeluted components, thus providing full structural information (i.e., sequence) not available from UV/VIS detectors.

LC-MS provides data acquisition and complete identification of hundreds of compounds in a single run. Mass spectrometry resolves the identity of compounds in complex biological mixtures (and also in combinatorial chemistry) through its ability to sequence online in real time. It is worth noting that retention time is not a criterion for sample identification. Structure verification (e.g., the amino acid sequence for proteins and peptides) is most important.

LC-MS CONNECTIONS

In modern LC-MS instruments, online connections are realized by direct coupling of the chromatographic column to an ESI source without flow split. In many cases, use of a flow split is desired or even necessary. Many commercial instruments are manufactured with a split to provide proper flow to the column. It should be noted

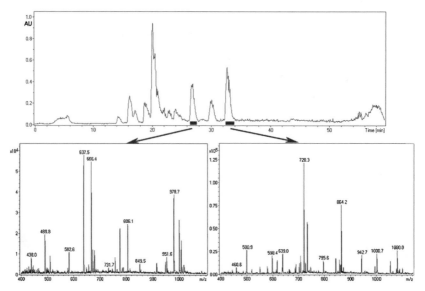

FIGURE 2   Comparison of a UV chromatogram (upper panel) with ESI-MS spectra acquired simultaneously during LC-MS analysis (lower left and right panels). Mass spectra from marked areas clearly show coelution of a few substances.

that such a configuration still utilizes a considerable amount of solvent, despite the fact that the final flow rate directed to the capillary column is in the nanoliter range.

Two setups are used: precolumn split and postcolumn split. Next, we provide a brief description of the principles and benefits of each.

Precolumn Split (Flow Splitter Located Between the Pumps and an Injector)   This type of connection, used in capillary HPLC systems, overcomes difficulties in maintaining nanoflow (Figure 3A). The flow splitter divides the flow from the pump or gradient former into two streams, one (at a higher flow rate) diverted to the waste

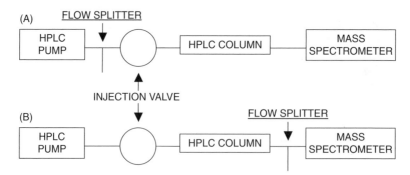

FIGURE 3   Location of flow splitters in LC-MS systems: (A) precolumn split; (B) postcolumn split.

TABLE 1   Advantages and Disadvantages of a Precolumn Split

Advantages	Disadvantages
Very good sensitivity	Entire sample consumed in the ion source
Capillary columns can be linked to the analytical HPLC system	System very sensitive to dead volumes
Very good gradient stability	Analysis time much longer than in conventional analytical HPLC

(or recycled) and the second directed to the injector. Achievable flow rates are in the range 50 to 500 nL/min. Such flow rates are easily accommodated by capillary columns of internal diameter 75 μm and length from 5 to 15 cm. The main advantages and disadvantages of this type of connection are listed in Table 1.

Postcolumn Split (Flow Splitter Located Between the Column Outlet and an ESI Source)   This type of connection is used with an analytical HPLC system and a column diameter 4.6 mm or larger, and eluted at a flow rate of 1 mL/min or higher (Figure 3B). Commercial ESI sources accept flow rates from 1 μL/min to 1 mL/min, but the upper limit often does not exceed 200 to 300 μL/min. Higher flow rates cause eluent overflow and may affect the ionization process and even lead to shortcuts in the source. The main advantage of such a connection is that only a small part (ca. 5%) of the sample is analyzed by the ESI. The remaining material may be collected and used for further experiments. The main drawback is lower sensitivity because only a fraction of the eluent enters the mass spectrometer. The most important advantages and disadvantages of this type of connection are given in Table 2.

CONSTRUCTION OF THE SPLIT

Splitters are available from several HPLC manufacturers. Also, inexpensive devices with regulated flow can be purchased, thus eliminating problems concerning calculations and adjustment of capillaries. However, if necessary, a homemade split can be prepared in the laboratory. The important components are the zero–dead volume tee and fused-silica capillaries of various diameters. As there is no common rule for flow calculation given, for example, differing solvent viscosities (gradient runs), a trial-and-error method is advised. In general, a capillary of larger diameter (75 μm

TABLE 2   Advantages and Disadvantages of a Postcolumn Split

Advantages	Disadvantages
Analytical columns can be used	Low sensitivity
Small fraction of the sample consumed in MS	Stable flow may be difficult to achieve
	Possible changes in flow rate during gradient separations

TABLE 3   HPLC Columns for LC-MS

Column Diameter	Flow Rate	Common Name	MS Compatibility
>100 mm	>50 mL/min	Preparative, industrial	Very limited, flow splitter needed or off-line
10–100 mm	2–50 mL/min	Semipreparative	Very limited, flow splitter needed or off-line
2–10 mm (4,6 mm)	0.5–2 mL/min	Analytic	With flow splitter or LC-MS online
0.8–2 mm	50–500 μL/min	Microanalytic (narrowbore, microbore)	Very good compatibility, LC-MS online
0.05–0.8 mm	0.05–50 μL/min	Capillary	Very good compatibility, nano-LC-MS online

i.d.) is linked to the system; a capillary of smaller diameter (20 μm) produces the appropriate backpressure and therefore is about three times longer.

COLUMNS FOR LC-MS ANALYSES

As mentioned earlier, mass spectrometers tolerate volatile buffers at low concentrations. Therefore, ideal solvents should contain water, methanol, and acetonitrile, often supplemented with trifluoroacetic (TFA) or formic acid (ca. 0.1%). In this chapter only basic criteria are discussed: (1) flow rate, (2) column length and diameter, (3) type of stationary phase, and type of eluent. Shown in Table 3 are column parameters that can be helpful when selecting an optimal LC system for a particular purpose.

The crucial factor for LC-MS systems, especially those working in the nanoflow range, is the reduction of all dead volumes. For example, working at a flow rate of 200 nL/min, a 1-μL dead volume will be filled after 5 minutes and may cause severe peak broadening and a delay in retention times. Peak broadening might sometimes be utilized to our benefit in the peak parking approach (see below).

PEAK PARKING

Peaks emerging from a capillary column are usually very sharp and about the last 15 to 30 s in a mass spectrometer. Each component needs to be analyzed in a full-scan mode and a sequence (i.e., MS/MS) should be identified. The total analysis time might be too long for a complete analysis. The simplest solution to this problem is the use of *peak parking*. As soon as a system detects the eluting component, the flow rate is reduced automatically, and thus the fraction can be kept in the mass spectrometer for a time sufficient for a complete analysis. This method causes peak broadening, but column resolution plays a somewhat less important role when MS is used as a detector. A certain level of experience is necessary, as the reduced flow can affect the stability of the spray.

STATIONARY PHASE

In modern HPLC we can distinguish several types of stationary phases used in proteomics research. Each type determines the nature of interactions among the column filling, eluent, and sample. The following properties are used most frequently:

1. *Reversed-phase chromatography* (RPC). Solid phase is based on nonpolar carbon chains from C_2 to C_{18}, with various modifications. Separation is based on the hydrophobicity of the compounds. Gradient elution is generally used. Small amounts of such acids as formic or trifluoroacetic are added to control the protonation level of side groups in molecules eluted and to improve peak shape. Recent developments include another type of stationary phase, based on polymeric support (monolithic columns and continuous beds). Such columns provide much lower backpressure, allowing higher flow rates, the stationary phase is bound covalently to capillary walls, and no protecting frits are necessary.

2. *Hydrophilic interaction* (HILIC). This technique is the inverse of RPC. The stationary phase is hydrophilic and the eluent is hydrophobic. Compounds are eluted in order of increasing hydrophilicity. Organic solvents of various polarities are used, with the addition of small amounts of salts. Alternatively, gradient elution is performed by decreasing the salt content on the column. Similar to ion exchangers, separation on HILIC material requires sample desalting.

3. *Separation based on the size of molecules* [size-exclusion chromatography (SEC), gel permeation chromatography (GPC), gel filtration chromatography (GFC), molecular sieving]. A sample is filtered through a molecular mesh made of a natural (dextran, polysaccharides, agarose) or synthetic (polyacrylamide) polymer. The retention time depends on the size of the molecules (or, more accurately, on their hydrodynamic index). Isocratic elution is used.

4. *Ion-exchange chromatography* (IEC). This type of chromatography is based on ion–ion interactions between sample components and stationary phase. This type of chromatography includes anion exchangers, where negatively charged ions are bound to the stationary phase, and cation exchangers, where positively charged ions interact with immobilized groups. Gradient elution from low to high salt concentration is used. For online LC-MS, involatile buffers must be replaced by other eluents to avoid ESI source clogging.

5. *Supercritical fluid chromatography* (SFC). Most commonly, liquefied carbon dioxide is used. The gas is in liquid form under a pressure of 20 to 30 atm and at ambient temperature. Its low viscosity and low surface tension under supercritical conditions allows for very effective separation of small molecules (e.g., metabolome analyses) on typical HPLC columns.

SUMMARY

Reversed-phase and cation-exchange columns are used most commonly in LC-MS. Reversed-phase materials (from C_2 to C_{18}) are most popular because they require volatile solvents (e.g., water, methanol, acetonitrile), fully compatible with ESI sources. RP HPLC is also a method of choice in the separation of peptides. In many cases, however, other materials should be used. For example, highly hydrophobic samples can be separated on an ion-exchange column. The column can be eluted with a NaCl gradient, but the sample needs to be desalted prior to analysis in a mass spectometer (see Chapter 7). Alternatively, volatile buffers can be used, based, for example, on ammonium acetate (pH 5.5) or ammonium bicarbonate (pH 8.5). Such mobile phases need to be supplemented with 5 to 10% acetonitrile or methanol, if possible, to maintain proper spray conditions in ESI sources. Usually, a combination of a cation exchanger, followed by a small desalting (trap) column and a capillary reversed-phase column are used for two-dimensional chromatography in proteomics (for details, see Chapter 3.2).

Acknowledgments

This work was supported by International Centre for Genetic Engineering and Biotechnology (ICGEB) grant CRP/POL05-02.

SELF-STUDY QUESTION

1. How can an ion-exchange chromatographic column be connected effectively (on line) to an electrospray ion source?
2. What are the advantages and disadvantages of two-dimensional over one-dimensional LC-MS?
3. What are major advantages of nano-LC-MS over an analytical LC-MS system?
4. Is it possible to recover or save native sample components during LC-MS analysis?

RECOMMENDED READING

Noga M., Sucharski F., Suder P., Silberring J. A practical guide to LC-MS troubleshooting. *J. Sep. Sci.* **30** (2007) 2179–2189.

Meiring H.D., Van Der Heeft E., ten Hove G.J., de Jong A.P.J.M. *J. Sep. Sci.* 25 (2002) 557–568.

Josic D., Clifton J.G. Use of monolithic supports in proteomics technology. *J. Chromatogr. A* 1144 (2007) 2–13.

Liao J.L., Li Y.M., Hjerten S. Continuous beds for microchromatography: reversed-phase chromatography. *Anal. Biochem.* 234 (1996) 27–30.

3.2

MULTIDIMENSIONAL LIQUID CHROMATOGRAPHY COMBINED WITH MASS SPECTROMETRY

ANNA DRABIK

Multidimensional protein identification technology (MudPIT) is devoted to efficient separation of complex mixtures of proteins and peptides. Although this definition has a much broader meaning, it is linked primarily to liquid separations. This approach, after adaptation, can also be utilized for separation of other molecules, such as small metabolites (metabolome).

Proteomic analysis of biological material is a very challenging task. Usually, the range of protein concentrations in a given sample may vary by 12 orders of magnitude. Analysis of a serum sample typically means identification of thousands of peptides. No single method is capable of managing such a task. Multidimensional separation can be an alternative to two-dimensional polyacrylamide gel electrophoresis (PAGE), keeping in mind that gel electrophoresis is not suitable for peptide separation, especially when proteins are also present in a sample.

A combination of two or more separation techniques characterized by different mechanisms of separation significantly improved the resolution and increased the number of proteins identified. Online linking of various types of columns contributes to shorter analysis time, reduces sample loss, and provides quantitative measurements. The general rule for construction of a multidimensional system is that to maintain good resolution of analysis each succeeding step should accommodate higher resolution than in the preceding step.

The first attempt to build a multidimensional instrument designed for protein separation was that of Smithies and Poulik in 1956. A two-dimensional electrophoresis

Proteomics: Introduction to Methods and Applications, Edited by Agnieszka Kraj and Jerzy Silberring
Copyright © 2008 John Wiley & Sons, Inc.

method was based on protein migration across polyacrylamide gel under denaturing conditions in the first dimension and in agarose in the second dimension.

Electrophoretic techniques have several restrictions, including limited sample capacity; the detection method, which depends on staining; unsatisfactory reproducibility; long analysis time; and difficulties with separation of "extreme" molecules or high- and low-molecular-mass compounds (very large complexes or short peptides). Despite the use of robots that can automatize spot picking and digestion, the method remains laborious and time consuming.

Alternative techniques are used, such as multidimensional chromatographic separation with high speed, automation, high throughput, and low risk of sample loss. Another option is that of isoelectric focusing in solution, followed by, for example, one-dimensional SDS-PAGE and LC-MS.

The resolving power of the combined methods depends on the resolution of each component present in the system:

$$\overline{R} = R_1 \times R_2 \times R_3 \times \cdots$$

The majority of setups utilize capillary systems to increase sensitivity. However, other combinations are also popular with off-line separations. Here we focus primarily on online capillary separations.

PRECOLUMNS AND SYSTEM PROTECTION

Capillary systems are very sensitive to solid contaminants, which can originate from the sample or eluents but also from solid-phase cartridges or homemade microcolumns filled with reversed-phase material. Often, vacuum evaporation or lyophilization results in insoluble residues. Contaminants can be eliminated using ultrafiltration, filtration, and/or precolumns that also protect analytical columns from obstruction. Furthermore, precolumns allow for sample preconcentration, salt removal, and initial sample separation. Precolumns usually contain the same type of stationary phase as that used in capillary analytical columns, the only difference being a larger diameter and shorter length.

SYSTEM SETUP

The rules for multidimensional system setup are:

- The columns should be connected beginning from the largest diameter to the smallest, located at the end of the system (i.e., closest to the ion source).
- The columns should be linked beginning at the lowest resolution.
- The columns should be positioned depending on the mobile phase; those eluted with volatile eluents should be located at the end to protect the ion source from salts.

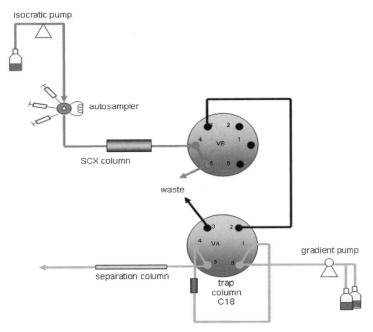

FIGURE 1   Coupling scheme for SCX and RP columns with a small trap cartridge in between. (*See insert for color representation of figure.*)

The latter requirement determines the order of column placement in the system. Ion exchangers will serve as a first separation step (salt gradient), followed by a reversed-phase column (organic solvents). When we consider the number of peptides in a given cell type released from all proteins upon trypsinization reaching (at least in theory) 1 million, it is obvious that the major challenge is to analyze peptide digest on a large scale, with high resolution and speed.

The two-dimensional coupling used most commonly in proteomic labs is a marriage of a strong cation exchanger (SCX) with a reversed-phase (RP) column. Separation on SCX maintains prefractionation (Figure 1). This step is usually performed with a stepwise gradient of salt, and volatile buffers should be used to minimize further problems. The discontinuous shape of the gradient ensures that only the desired number of fractions will be eluted. This, in turn, provides an optimal number of runs on the next column. Yet another feature of stepwise elution is that after each segment of the gradient, the system is stopped and the eluted fraction is diverted onto the trap column. Therefore, use of a linear gradient is of less importance.

OFF-LINE SEPARATIONS

It is not uncommon for researchers to try to apply a "trendy" technique despite difficulties associated with the methodology. In many cases it is much simpler to use multidimensional separations in an off-line setup. Usually, the SCX column is disconnected from the system and its dimensions are larger (2 to 3 mm i.d.), which provides a higher capacity, and much more material can be fractionated. Here either linear or discontinuous gradients can be used. Fractions eluted from the column are collected, desalted, and analyzed online with an LC-MS system.

SEPARATION AND IDENTIFICATION

Peptides are eluted as shown in Figure 2. The volatile buffer used in this method is more "friendly" to the trap column and does not clog it to the same extent as does NaCl. As can be seen, the first dimension accomplishes rough separation into several fractions, each of them still contains many components. After desalting on a small trap column, each fraction will be separated on a reversed-phase column in the second dimension. Purified components (peptides) are then analyzed in the mass spectrometer (LC-MS/MS) and identified with the help of protein databases (Chapter 7.5).

Figure 3 shows consecutive phases of protein identification. One of the fractions eluted from the SCX column has been desalted and preconcentrated on the trap column, and the components were separated on a reversed-phase capillary column linked to a nanospray interface. Full-scan spectra were taken from all peptides (above

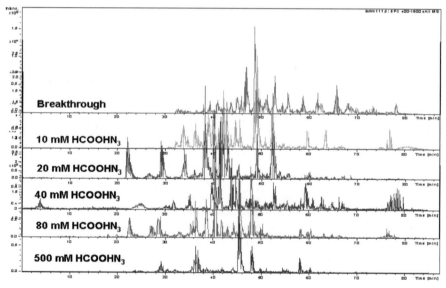

FIGURE 2   Chromatograms presenting peptides eluted from an SCX column (first dimension) after subsequent separation on a reversed-phase capillary column. Concentrations of the salt segments (discontinuous gradient) are also shown. (*See insert for color representation of figure.*)

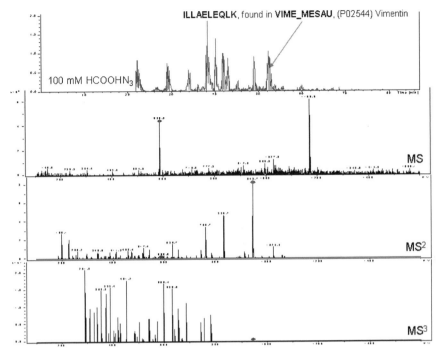

FIGURE 3   Identification of protein based on sequential separation and analysis of a peptide tag using an SCX/RP-MS^3 setup.

a certain intensity threshold), followed by MS/MS and MS^3. Protein identification was carried out with the help of sequence tags and a protein database.

MULTIDIMENSIONAL SYSTEMS

Another arrangement utilizes a three-dimensional system, whose configuration is presented in Figure 4. This type of connection allows for the identification of an even larger number of proteins. Evaluation of RP-MS/MS, SCX/RP-MS/MS, and RP/SCX/RP-MS/MS potentials gave the following results: 147 proteins identified using a one-dimensional technique; 314 by a two-dimensional technique; and finally, 413 proteins revealed based on a three-dimensional method (McDonald et al., 2002). At present, the performance of such systems is far better, but their capability, resolution, and time of analysis are similar. The choice of a desired system is always a compromise among costs, effort, technical difficulties, and outcome (for additional information, consult the recommended reading list).

Naturally, as the complexity of a system increases, the analysis time also increases as the number of steps increases: the average time for a one-dimensional method is 1 hour, for a two-dimensional method is 12 hours, and for a three-dimensional method is 100 hours.

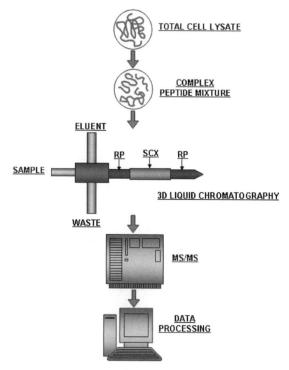

FIGURE 4   Three-dimensional arrangement of SCX and RP columns used for highly efficient separation.

Multidimensional separation involving affinity chromatography (AC) makes it possible to isolate compounds with characteristic functional groups. This is the most selective purification method available and is often used for the specific identification of glycoproteins or phosphoproteins.

Acknowledgments

This work was supported by International Centre for Genetic Engineering and Biotechnology (ICGEB) grant CRP/POL05-02 and ENACT (European Network for the Identification and Validation of Antigens and Biomarkers in Cancer and Their Application in Clinical Tumor Immunology) contract 503306.

SELF-STUDY QUESTIONS

1. Consider a two-dimensional LC-MS setup with a size-exclusion column as a first dimension and a reversed-phase capillary column as a second dimension. Discuss advantages and disadvantages of this method for the separation of

proteins and peptides and compare with the "classical" approach utilizing SCX and RP columns.

2. How can 5 μL of sample solution be introduced into a nano-LC-MS system in view of the eventual diffusion of the material injected?

RECOMMENDED READING

Issaq H.J. Application of separation technologies to proteomics research. *Adv. Prot. Chem.* 65 (2003) 249–269.

Issaq H.J., Chan K.C., Janini G.M., Conrads T.P., Veenstra T.D. Multidimensional separation of peptides for effective proteomic analysis. *J. Chromatogr.* 817 (2005) 35–47.

Link A.J. Multidimensional peptide separations in proteomics. *Trends Biotechnol.* 20 (2002) 8–13.

McDonald W.H., Ohi R., Miyamoto D.T., Mitchison T.J., Yates J.R. Comparison of three directly coupled HPLC MS/MS strategies for identification of proteins from complex mixtures: single-dimension LC-MS/MS, 2-phase MudPIT, and 3-phase MudPIT. *Int. J. Mass Spectrom.* 219 (2002) 245–251.

Neverova I., Van Eyk J.E. Role of chromatographic techniques in proteomic analysis. *J. Chromatogr. B* 815 (2005) 51.

Noga M., Sucharski F., Suder P., Silberring J. A practical guide to nano-LC-MS troubleshooting. *J. Sep. Sci.* 30 (2007) 2179–2189.

Smithies O., Poulik M.D. Two-dimensional electrophoresis of serum proteins. *Nature* 177 (1956) 1033.

Stroink T., Ortiz M.C., Bult A., Lingeman H., de Jong G.J., Underberg W.J. On-line multidimensional liquid chromatography and capillary electrophoresis systems for peptides and proteins. *J. Chromatogr. B* 817 (2005) 49–66.

Wang H., Hanash S. Multi-dimensional liquid phase–based separations in proteomics. *J. Chromatogr. B* 787 (2003) 11–18.

4

ELECTROPHORETIC SEPARATION METHODS

4.1

ONE-DIMENSIONAL GEL ELECTROPHORESIS

Anna Drabik and Piotr Laidler

Founded almost 60 years ago, protein electrophoresis remains the most effective way to resolve a complex mixture of proteins. Electrophoresis can be used as a preparative tool to obtain a pure (or almost pure) protein sample, or as an analytical tool to provide information on the mass, charge, purity, or presence of proteins, peptides, and nucleic acids.

SAMPLE PREPARATION

Different proteomic approaches use a variety of techniques. All of them, in some form, involve protein isolation, separation, quantitation, and identification. Prior to fractionation, proteins need to be denatured, disaggregated, reduced, alkylated, and solubilized, to achieve complete disruption of molecular interactions (i.e., down to the first-order structure). A number of protocols have been published, and Internet links are listed in Appendix C. Some general recommendations can, however, be made:

- Sample preparation should be as simple as possible.
- Protein modifications should be minimized.
- Proteolytic enzymes must be inactivated.

As with any other experimental research, the quality of the results depends entirely on the quality of the preliminary material. Three basic stages during sample

Proteomics: Introduction to Methods and Applications, Edited by Agnieszka Kraj and Jerzy Silberring
Copyright © 2008 John Wiley & Sons, Inc.

preparation are (1) cell disruption, (2) inactivation or removal of interfering substances, and (3) protein solubilization.

Cell disruption can be achieved by using one of the methods listed or by a combination: osmotic lysis, detergent lysis, enzymatic lysis, freeze–thawing, sonication, grinding with liquid nitrogen, high pressure, and homogenization with glass beads or rotating blades.

Protease inactivation should be carried out to prevent the formation of artifactual spots, proteolysis, and losses of high-molecular-mass proteins, by adding protease inhibitors (i.e., an inhibitory cocktail). Also, utilization of an ice-cold trichloroacetic acid (TCA), high- or low-pH, boiling sample in sodium dodecyl sulfate (SDS) buffer, denaturation in boiling water or 1 M acetic acid, microwave irradiation, and/or organic solvents can cause inactivation of proteases and the recovery of higher quantities of biological material. The addition of chaotropes (e.g., thiourea, urea) causes hydrogen bond disruption as well as hydrophilic interactions between protein molecules.

Protein solubilization using detergents prevents hydrophobic interactions between hydrophobic protein domains, along with protein loss due to aggregation and precipitation. SDS is an anionic detergent that denatures proteins by "wrapping around" the polypeptide backbone. It binds specifically to proteins in a mass ratio of 1.4 : 1. SDS creates a negative charge in proportion to the polypeptide chain length with equal charge or charge densities per unit length. In denaturing SDS-PAGE separations, therefore, migration is determined not by the intrinsic electrical charge of the polypeptide, but by molecular mass. Other commonly used detergents are Nonidet P-40, t-octylphenoxypolyethoxyethanol (Triton X-100), t-octylphenoxypolyethoxyethanol (Triton X-114), cetyltrimethylammonium bromide (CTAB), polyoxyethylen[23]lauryl ether (BRIJ 35), polyoxyethylene[20]cetyl ether (BRIJ 58), polyoxyethylenesorbitan monolaurate (Tween 20), polyoxyethylenesorbitan monooleate (Tween 80), n-octyl-β-D-glucopyranoside (octyl-β-glucoside), (3-[3-chloamidopropyl]dimethylammonio)-1-propanesulfonate (CHAPS), (3-[3-cholamidopropyl]dimethylammonio)-2-hydroxy-1-propanesulfonate (CHAPSO).

Reduction and avoidance of reoxidation of the disulfide bond is also a significant step in the sample preparation protocol. Reducing agents are required for cleavage of intra- and intermolecular disulfide bonds and achieving total protein unfolding. Typically used reduction agents are dithiothreitol (DTT), dithioerythriol (DTE), tributylphosphine (TBP), tris(2-carboxyethyl)phosphine (TCEP), and β-mercaptoethanol (BME). To protect the free cysteine groups' iodoacetamide, an alkylation agent should be added.

Alkylation causes blocking of the most reactive groups in peptides and proteins—SH thiol groups attached to primary carbon atoms. There are many other methods, but carbamidomethylation with iodoacetamide and pyridylethylation are the most popular.

High salt concentration that is a consequence of its presence in bodily fluids and tissues may interfere with electrophoretic separation, because of its high electrical current. Typically, salt elimination methods used are microdialysis, ultrafiltration, precipitation, solid-phase extraction, spin filters, and the use of columns. Moreover, exclusion of the most abundant proteins (e.g., albumin, immunoglobulins) in a sample improves the electrophoretic resolution.

SEPARATION MECHANISM

After electric current is applied, molecules placed in particular wells migrate, and the distance they move through depends on pore size, molecular weight, and total charge. Larger molecules are slowed down by the gel pores, and in effect they are localized on the top of the slab. Migration is directly proportional to the velocity of the ion and inversely proportional to the electric current force. *Electrophoretic mobility* is the constant of proportionality between the charge of the ion and the friction coefficient:

$$\mu = \frac{V}{E} = \frac{Z}{f}$$

where μ is the electrophoretic mobility, V the velocity of the ion, E the strength of the electric field, Z the total molecular charge, and f is a friction coefficient. The electrophoretic mobility also depends on the shape of molecules and their hydrophobicity and viscosity, the ion force, and the buffer temperature:

$$f = 6\pi\eta r$$

where η represents the viscosity and r the hydrodynamic radius. Most molecules are not spherical, but this equation defines an effective hydrodynamic radius of a molecule as if it were a sphere, based on its observed mobility.

SDS-PAGE

Sodium dodecyl sulfate polyacrylamide gel electrophoresis (SDS-PAGE), the most widely used electrophoretic technique (Figure 1), separates proteins primarily by mass but also by shape and size influence on migration. Many proteomic applications is based on one-dimensional electrophoresis, because it is simple to perform, reproducible, and can be used to separation proteins that have a wide range of molecular masses (0.5 to 300 kDa). Smaller proteins progress through gel faster than do larger proteins. Generally, a protein mixture migrates through polyacrylamide gel in the presence of SDS, a negatively charged detergent, to unfold proteins and break the interaction with other molecules. When electrical current is applied, the proteins migrate toward positive electrodes down through the gel matrix, creating lanes of protein bands.

Polyacrylamide gel is created from the polymerization of two toxic compounds, acrylamide and N, N^1-methylenebisacrylamide (Figure 2). Bisacrylamide is a cross-linking agent. Polymerization starts with the addition of ammonium persulfate along with either β-dimethylaminopropionitrile (DMAP) or N, N, N^1,

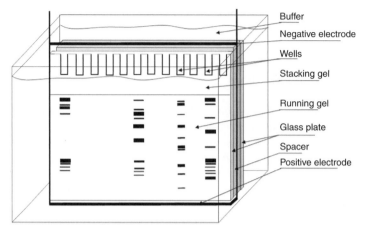

FIGURE 1   Apparatus for the SDS-PAGE separation technique.

N^1-tetramethylethylenediamine (TEMED). The gels are neutral, hydrophilic, three-dimensional networks build with long hydrocarbons cross-linked by methylene groups.

The separation of molecules is determined by the size of the pores formed inside the gel. The pore size depends on two factors, the total amount of acrylamide present and the amount of a cross-linker:

$$\%T = (\text{acrylamide} + \text{bisacrylamide}) \, (\text{g})$$
$$\%C = \frac{(\text{bisacrylamide}) \, (\text{g}) \times 100}{(\text{acrylamide} + \text{bisacrylamide}) \, (\text{g})}$$

As the total amount of acrylamide increases, the pore size decreases. The smallest pores are obtained with the 5% cross-linker concentration, while any increase or decrease of its concentration increases the pore size. Gels are considered as percent solutions and have two necessary parameters: the total acrylamide and the bisacrylamide amount. Proteins with molecular mass ranging from 10 to 1000 kDa might be separated with 7.5 g of acrylamide and 5% of bisacrylamide, while proteins with higher masses require lower acrylamide gel concentrations. Gels up to 30% have been

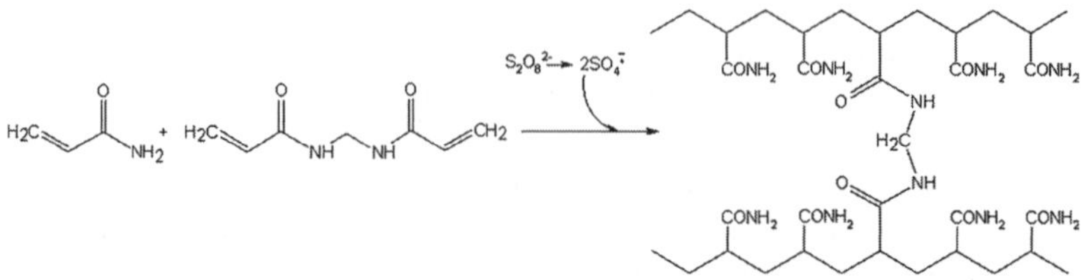

FIGURE 2   Polyacrylamide polymerization.

TABLE 1   Gel Percentage for Protein Separation at Various
Molecular Masses

% Gel	Molecular Mass Range (kDa)
7	50–500
10	20–300
12	10–200
15	3–100

used to separate small polypeptides. The higher the gel concentration, the smaller the pore size and the better the separation of smaller molecules (Table 1).

The polymerization time depends on temperature and catalyzer concentration. At room temperature, acrylamide usually polymerizes after 20 to 30 minutes. However, complete net creation takes less than 12 hours. Oxygen is an inhibitor of this reaction.

The original application of gels as a separating medium concerned using a single gel with a homogeneous pH. This method has been replaced by discontinuous, multiple-gel systems, where a separating gel is improved using a stacking gel and optional sample gel (Figure 3). Each gel segment may have special concentrations of the same support media or may be completely unlike. The primary difference lies in the way that molecules separate when they penetrate the following gel. The sample gel causes the concentration of proteins into a small zone in the stacking gel before entering the separating gel. When the proteins are concentrated, they continue their motion into the separating gel in narrow bands. Afterward, the bands are separated from each other on a discontinuous pH gel.

In a gradient procedure, gel is divided into three segments, known as the *separating* or *running gel*, the *stacking gel*, and the *sample gel*. The sample gel may be replaced by a dense nonconvective medium such as sucrose.

When electrodes are placed in such manner that the upper bath is a cathode while the lower bath is anodic, and anions are heading toward the anode, the system is

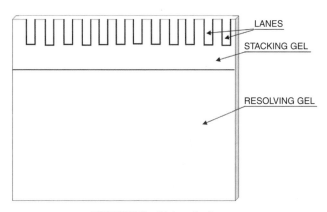

FIGURE 3   Slab-gel scheme.

called an *anionic system*. Movement in the opposite direction, with cation motion toward the cathode, signifies a *cationic system*.

Slab gels are considered for multiple sample separation in one run. The size and number of lanes can differ according to one's need. The reproducibility of results increases since when samples are run in the same medium, there is less likelihood of sample variation due to minor changes in gel structure. Slab gels comprise a technique designed for Western blot and radiographic analysis. The accessibility and relatively low cost of salable gel equipment are among the benefits of this method and have increased the utilization of slab-gel systems.

AGAROSE GELS

Although acrylamide gels have become the standard for protein separation, they are less appropriate for high-molecular-mass nucleic acids. At the concentration required, the acrylamide is reduced to a level at which it remains liquid. The addition of linear polysaccharide agarose to the low concentration of acrylamide enables the formation of gel molecular masses up to 1 million daltons that can be separated. Agarose gel is functional principally for the separation of large DNA sequences. Therefore, agarose–acrylamide gels are used primarily for the determination of gene maps.

NATIVE PAGE

Electrophoretic technique in a nondenaturing environment, also called *native PAGE*, separates proteins according to their charge and hydrodynamic size. In native PAGE, migration occurs, as proteins possess a negative charge. The higher the negative charge concentration, the more rapidly a protein will travel. There are no denaturants present in native PAGE, so subunit interactions within separated proteins retain their enzymatic activity. Thus, native PAGE may be used for research into active proteins. Proteins might be recovered from a native gel through passive diffusion or electroelution. Users should remember that it is essential to keep the device cool, to minimize the effects of denaturation and proteolysis.

STAINING METHODS

After electrophoretic separation, proteins must be visualized in some manner. A number of staining techniques exist. Selection of the proper detection method is extremely important, because the proteomic approach often relies on measuring quantitative changes in expression level. The detection limit should be as low as possible (however, it does not need to be lower than the MS analysis detection limit) with an optimal signal-to-noise ratio. The process should be easy and fast, nontoxic, cheap, and MS compatible. The dye should have a wide linear relationship between the quantity of protein and the staining intensity.

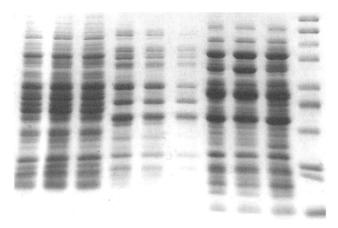

FIGURE 4   SDS-PAGE scan after biological sample separation and CBB staining.

Coomassie Brilliant Blue

Organic dye Coomassie Brilliant Blue (CBB) was used for protein detection at the beginning of the 1960, and is still the most commonly used method for protein detection in polyacrylamide gels. There are two types of CBB tints that produce blue bands on a clear background with a sensitivity of 8 to 10 ng: greenish, G-250, and reddish, R-250 (although R-250 presents a bit lower sensitivity). In an acidic environment, particles of dye bind to amino groups of the proteins by electrostatic and hydrophobic interactions. The proteins possess a higher affinity for CBB than the polyacrylamide, which is why background destaining is possible. The CBB visualization technique is easy to use, highly reproducible (a coefficient of variation below 10%), linear up to three orders of magnitude, and MS compatible.

CBB dissolved in alcohol provides lower reproducibility than that of a colloidal solution (CBB G-250 dissolved in TCA or phosphoric acid), because proteins such as collagen destain faster than does polyacrylamide gel. Another advantage of using colloidal CBB is that there is no requirement for a destaining step. A fast, hot CBB staining method is less time consuming, but the sensitivity is not as high as for a colloidal CBB dye. The "blue silver" technique was reported to be more sensitive (with a limit of detection around 1 ng), but it requires higher acid and dye concentrations.

Summarizing, the CBB staining method shows quantitative binding to proteins (Figure 4), good reproducibility, low costs, and when it detects a protein, the user can be sure that it can be identified using a mass spectrometer.

Silver Staining

Silver staining is widely used, despite the multistep protocol required, because of its sensitivity (below 1 ng of protein per spot). Detection is based on silver precipitation on the surface of the gel, which is represented by dark brown spots on a bright background. Depending on the conditions, there are two types of silver staining: alkaline and acidic, although ammoniacal staining is reported to be more sensitive. Silver staining provides very high reproducibility and allows for quantitative determination of proteins. However, it represents a very narrow range, so the quantitation of highly abundant proteins is not reliable, because spots often show a yellow center which during image analysis looks like a volcano crater. Depending on the protocol, as little as 100 pg of protein can be detected using this technique. In general, silver staining shows a narrow dynamic range, relatively low cost, and high sensitivity.

Amido Black, Copper, and Fast Green Staining

Amido black, copper, and fast green stain, although representing sensitive high-speed methods, were never well accepted for proteomic applications.

Zinc Salt, SDS, and Imidazole Staining

Zinc salt, SDS, and imidazole mixture staining is based on the visualization of colorless protein spots placed on a dark background. This method does not allow quantitative analysis of proteins because it does not modify the protein but rather, the background (negative staining method). The peptide yield after in-gel digestion is reported to be higher than in other techniques mentioned. Despite those advantages, this detection method was never popular among proteomic researchers.

Fluorescence Staining

Fluorescence staining shows a very wide dynamic range (as much as four orders of magnitude) and high reproducibility. Fluorescence dyes can be used before or after electrophoretic separation. There are number of noncovalent fluorescent dyes, mostly SYPRO dyes. SYPRO Red and Orange bind to the detergent particles surrounding proteins in gel. SYPRO Ruby is a ruthenium chelate that binds to the basic amino acids in proteins. The visualization protocol involves only one step, it is quite simple, and it can be a high-throughput, large-scale application. Depending on conditions, 1 to 10 ng of protein can be detected. This is a reversible process. Excitation can be achieved with either a laser or a UV scanner. A detection limit of 0.5 to 5 ng was reported, and the linear dynamic range is three orders of magnitude. Fluorescence dyes allow for detection of glycoproteins (Pro-Q-Emerald), phosphoproteins (Pro-Q-Diamond), or metabolic labeling with ^{35}S. Epicocconone, a substance that has been isolated from fungus, appeared to be useful as a fluorescent stain in a technique called Lighting Fast. Deep Purple Total Protein Stain (that is the commercial name of the epicocconone) is very sensitive, with a linear dynamic range of over four orders of magnitude.

Prelabeling of proteins with a fluorescent dye prior to SDS-PAGE is also used. Combination of at least two fluorophores in one gel allows for differential imaging [difference gel electrophoresis (DIGE)]. DIGE permits the separation of up to three samples in one run. Each should be labeled with one of the fluorescence Cy dyes (Cy3, Cy5, or Cy2), then mixed and separated electrophoretically. This method yields quantitative and qualitative results of high accuracy, because of the elimination of gel-to-gel variation. A small number of gels are necessary for statistical analysis during comparison of healthy versus pathological states (e.g., normal and cancerous tissues).

Isotope Labeling

Isotope labeling is rarely employed in proteomic research. Radioactive labeling using ^{125}I, ^{14}C, ^{35}S, ^{32}P, or ^3H due to problems with reproducibility and sensitivity (a wet gel sheet disrupts the radioactivity signal). The very high sensitivity of fluorography is not adept for proteomics, because spots with low emission do not contain enough protein for MS analysis. Among isotope-labeling methods, only stable isotope labeling (using ^{14}N vs. ^{15}N or ^{12}C vs. ^{13}C) has been used successfully for two-dimensional electrophoresis. Once proteins have been detected, image analysis is required. Various companies offer software for digital image examination. In Table 2, we summarize various staining techniques together with their features.

TABLE 2   Comparison of Staining Techniques

Staining Method	Advantages	Disadvantages
Coomassie	MS compatible Linearity range up to three orders of magnitude (colloidal solution) Low cost Good reproducibility	Detection limit 8 to 10 ng
Silver ammonia	Detection limit below 1 ng High reproducibility	Linearity range one order of magnitude Multistep protocol
Zinc-imidazol-SDS	MS compatible High sensitivity Detection limit 1 to 10 ng	Negative staining makes quantitative analysis impossible
Fluorescent dye	MS compatible High reproducibility Detection limit around 1 ng Reversible process Simple procedure Linearity range up to three orders of magnitude Possibility of PTMs' analysis	Requirement to use an ultraviolet or laser scanner
Isotope	High sensitivity Detection limit below 300 pg	Low reproducibility

LIMITATIONS OF ELECTROPHORETIC SEPARATION

Electrophoretic separation is a slow and labor-intensive procedure that is not easily automated. One-dimensional electrophoresis has limited resolution (around 50 proteins), which might cause overlapping of closely separate bands. The presence of keratins gives a false image of the sample contents. In addition, acrylamide is a toxic substance that produces background peaks on an MS spectrum.

Acknowledgments

This work was supported by International Centre for Genetic Engineering and Biotechnology (ICGEB) grant CRP/POL05-02 and ENACT (European Network for the Identification and Validation of Antigens and Biomarkers in Cancer and Their Application in Clinical Tumor Immunology) contract 503306.

SELF-STUDY QUESTIONS

1. A protein consisting of a single polypeptide chain is subjected to SDS-PAGE under the usual conditions. The same protein is also run in the absence of a reducing agent. The mobility of the protein in the SDS medium is the same in both cases. What conclusion might you draw from these results?
2. A protein with a molecular mass of 20,000 Da moves 3.0 cm in SDS-PAGE; a protein with 40,000 Da moves 2.0 cm under the same conditions. What migration distance is expected for a protein having a molecular mass of 30,000 Da?
3. Proteins A and B, with molecular masses of 24,500 and 42,300 Da, move 4.6 and 1.3 cm, respectively, in SDS-PAGE. What is the molecular mass of protein C, which moves 2.8 cm in the same gel?

RECOMMENDED READING

Gallagher S.R. One-dimensional SDS gel electrophoresis of proteins. *Curr. Protocols Mol. Biol.* 10(2A) (1999) 1–34.

Gharahdaghi F., Weinberg C.R., Meagher D.A., Imai B.S., Mische S.M. Mass spectrometric identification of proteins from silver-stained polyacrylamide gel: a method for the removal of silver ions to enhance sensitivity. *Electrophoresis* 20 (1999) 601–605.

Hamdan M., Bordini E., Galvani M., Righetti P.G. Protein alkylation by acrylamide, its N-substituted derivatives and cross-linkers and its relevance to proteomics: a matrix assisted laser desorption/ionization–time of flight–mass spectrometry study. *Electrophoresis* 22 (2001) 1633–1644.

Miller I., Crawford J., Gianazza E. Protein stains for proteomic applications: Which, when, why? *Proteomics* 6 (2006) 5385–5408.

Westermeier R., Marouga R. Protein detection methods in proteomics research. *Biosci. Rep.* 25 (2005) 19–32.

4.2

TWO-DIMENSIONAL GEL ELECTROPHORESIS

ANNA BIERCZYNSKA-KRZYSIK AND GERT LUBEC

Two-dimensional gel electrophoresis (2-DE), described by O'Farrell and Klose in 1975, remains one of a core tools for expression proteomics today. Substantial improvements introduced in past years, render two-dimensional electrophoresis capable of resolving up to several thousands of proteins simultaneously on a single gel. The technique is exceptional in its ability to indicate proteins' posttranslational modifications and isoforms. Other applications involve purity examination, detection of disease markers, analysis of cell differentiation, cancer research, and drug discovery.

Two-dimensional gel electrophoresis sorts proteins according to two independent properties, in two major steps (Figure 1), occurring in the presence of polyacrylamide gels (the most common approach[1]):

1. *First-dimension separation:* isoelectric focusing (IEF); proteins are separated on the basis of their isoelectric points (pIs).
2. *Second-dimension separation:* sodium dodecyl sulfate–polyacrylamide gel electrophoresis (SDS-PAGE); proteins are separated by their molecular weight.

Proteins separated via IEF and SDS-PAGE remain invisible on the gel until stained. An exemplary protein profile resulting from two-dimensional electrophoretic separation of brain proteins stained with Coomassie Brilliant Blue (CBB) is depicted in Figure 2. Spots visible on the gel represent individual proteins or a mixture of

---

[1]To separate high-molecular-mass proteins, agarose gels are often used instead of polyacrylamide gels.

Proteomics: Introduction to Methods and Applications, Edited by Agnieszka Kraj and Jerzy Silberring
Copyright © 2008 John Wiley & Sons, Inc.

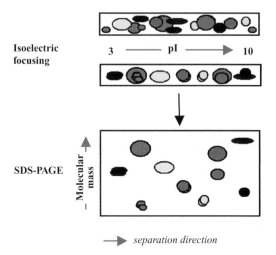

Isoelectric focusing

3 ——— pI ———▶ 10

SDS-PAGE

Molecular mass

——▶ *separation direction*

FIGURE 1 Two-dimensional gel electrophoresis scheme: proteins sorted according to their net charge during an isoelectric focusing step are subsequently separated by molecular weight via SDS-PAGE, perpendicular to the IEF direction. Spots in color represent proteins of various shapes, sizes, and pI values. (*See insert for color representation of figure.*)

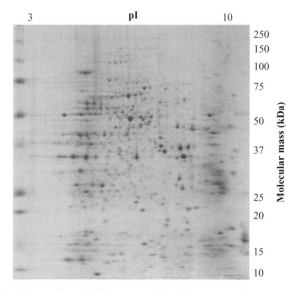

FIGURE 2 Protein profile of 800-μg proteins from rat striatum separated via two-dimensional gel electrophoresis (personal results). Protein standards of known molecular mass are visible on the right side of the gel. Protein pIs may be estimated by linear interpolation relative to protein standards of known pI.

several proteins.[2] Protein spots can be further excised and digested proteolytically or chemically. Peptides resulting from such cleavage constitute a fingerprint unique for each protein, and thus lead to protein identification with the use of mass spectrometry and bioinformatics.

FIRST-DIMENSION SEPARATION: IEF

IEF, the first step in two-dimensional gel electrophoresis, enables separation of proteins due to their isoelectric points (pIs). *pI* is defined as the pH at which a protein carries no net electrical charge and cannot migrate in an electric field.

At pH below a protein's pI, the molecule remains positively charged, whereas if the pH of the surrounding exceeds the pI value, the protein carries a net negative charge. These observations constitute the clue for use of an IEF separation technique.

The isoelectric point is determined by the number and types of charged groups in amino acids building a protein, which, depending on the surrounding pH, contribute to the total protein charge (Figure 3). For most proteins, the pI value oscillates between 4 and 8. It may be calculated from the p*K* values of a molecule:

$$pI = \frac{pK_1 + pK_2}{2}$$

In an electric field, proteins positioned in a medium with a pH gradient migrate toward an electrode of opposite charge. During migration, proteins gain or lose protons (depending on the initial protein charge); thus, their net charge and consequently their mobility decrease (Figure 4).

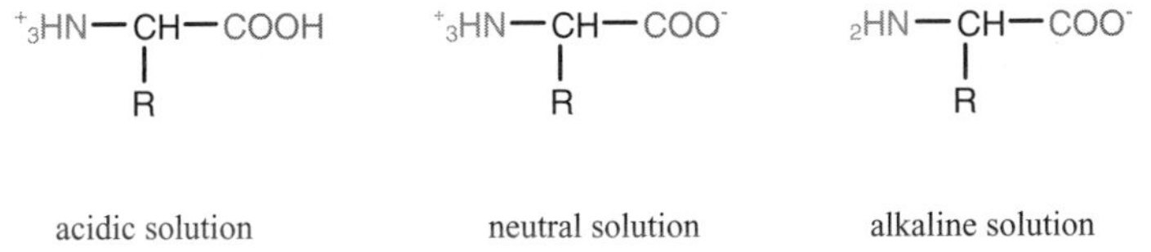

$$^+_3HN—CH—COOH \qquad ^+_3HN—CH—COO^- \qquad _2HN—CH—COO^-$$

$$\qquad\qquad | \qquad\qquad\qquad\qquad | \qquad\qquad\qquad\qquad |$$

$$\qquad\qquad R \qquad\qquad\qquad\qquad R \qquad\qquad\qquad\qquad R$$

acidic solution   neutral solution   alkaline solution

FIGURE 3 Predominant pH-dependent ionization states of an amino acid.

---

[2] Due to the insufficient resolving power of the two-dimensional gel electrophoresis technique, proteins of similar molecular mass and pI value often migrate together, and thus spots representing them on the gel overlap.

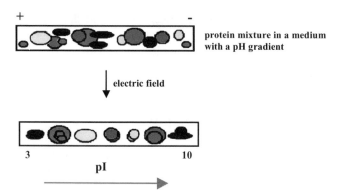

FIGURE 4 Isoelectric focusing: migration of proteins in an electric field. (*See insert for color representation of figure.*)

Proteins of positive net charge migrate, through a pH gradient, toward a negatively charged cathode, gradually decreasing their positive charge until they reach a pH value at which they have no net charge. Similarly, proteins of negative net charge migrate toward a positively charged anode, gradually decreasing their negative charge until they reach their pI value. Once all molecules are positioned at their pI, the focusing step is finished.

Even though the size of proteins with identical pI values may vary significantly, the molecules are focused at similar positions due to protein denaturation prior to experiments and to the lower resolution of such gels.

In contrast to one-dimensional gel electrophoresis, where proteins keep migrating through a polyacrylamide medium as long as an electric field is applied, proteins separated via IEF retain positions once achieved.

To enable an IEF process, strips with an immobilized pH gradient (IPG) are exploited. A pH gradient is generated by acrylamide derivatives, immobilines copolymerized with an acrylamide matrix, on a plastic sheet. IPG strips allow generation of pH gradients of any desired range between pH 3 and 12. Most proteins are displayed through the use of broad- and medium-range strips of pH 3–10 and 4–7 or 4–9, respectively. Narrowing the pH range with the use of overlapping IPGs (e.g., IPG 4.0–5.0, 4.5–5.5, and 5.0–6.0) and extension of separation distances make possible optimal resolution within a desired area (zoom-in gels). Since proteins of pI other than those chosen do not appear on a gel, more proteins may be loaded on a medium-range strip. As a result, even minor components of a complex sample are to be detected.

Separation in the first dimension includes:

- IPG strip rehydration
- Sample application
- Isoelectric focusing

Generally, proteins to be separated are introduced and simultaneously distributed over the length of a strip with a rehydration buffer (commercially available strips are in dried form). Rehydration, sample absorption, and IEF are usually performed in individual ceramic strip holders laid on the cooled base of electrophoresis apparatus, which allows IEF to be performed at high voltage. Another technique, cup loading, requires strip rehydration prior to sample application. Although beneficial for samples containing high concentrations of very high molecular mass, alkaline, or very hydrophobic proteins (in-gel absorption of these in the presence of rehydration buffer is troublesome; proteins tend to stay in solution), cup loading should be avoided, due to the occurrence of artifacts.

IEF is performed by gradually increasing the voltage across IPG strips and maintaining it for a consistent, empirically determined number of volt-hours (IEF is usually completed after 12 hours when ion movement is minor—reflected by a very low current). Low voltage, applied during sample loading, enables more effective entrance of high-molecular-mass proteins into the IPG strip. Moreover, voltage restriction to 50 to 100 V during the first hours of IEF allows for direct desalting of the sample in the IPG gel. During the step, salt ions responsible for initially high conductivity migrate toward electrodes. As a result, conductivity decreases and high voltages (up to 8000 V) may be used. The temperature should be kept constant, usually about 20°C, due to its effect on final protein patterns. Its elevation may lead to carbamylation of proteins (Chapter 8), whereas a decrease may cause precipitation of several buffer components (e.g., urea).

Once isoelectric focusing has been completed, IPG strips are equilibrated in two SDS buffers, enabling protein reduction and alkylation, respectively. The presence of SDS assures hydrogen bond disruption, blockage of hydrophobic interactions, and partial unfolding of the proteins. Moreover, as explained in Chapter 3.1, the anionic detergent contributes a strong negative charge that masks the intrinsic charge of proteins and alters their native conformation. Thus, electrophoresis in SDS-polyacrylamide gels permits separation of molecules almost exclusively on the basis of molecular mass, regardless of their pI values or shapes. Addition of dithiothreitol (DTT) in a first equilibration buffer enables complete unfolding of proteins by reducing cysteine disulfide bridges. Iodoacetamide implemented in the second buffer chemically modifies thiol groups in proteins, preventing their oxidation and re-formation of high-molecular-mass aggregates.

Next, IPG strips are applied to SDS-polyacrylamide gels. Alternatively, before the equilibration step, strips may be stored frozen at –40 to –80°C. Multiple two-dimensional gels may be run simultaneously. Electrophoresis chambers usually hold from 10 to 12 gels. A gel strip is overlaid horizontally on top of SDS-polyacrylamide gel and a mixture of protein molecular mass standards is usually applied on filter paper and placed next to a strip. Sealing a strip with warm agarose removes air bubbles between a strip and a gel surface (bubbles may cause uneven spot distribution). Cooled and concentrated agarose ensures good contact between a strip and a gel surface and prevents a strip from changing its position.

SECOND-DIMENSION SEPARATION: SDS-PAGE

The basics of SDS-PAGE, following an IEF step and exploited in comparative proteomic studies, is described in Chapter 3.1. The only difference is a way of sample loading onto SDS gel: In one-dimensional gel electrophoresis, solutions containing proteins of interest are introduced directly into separate wells, whereas the two-dimensional form exploits an IPG strip as a protein source.

SDS-PAGE occurs perpendicular to IEF. Depending on the gel size and concentration and the applied voltage, electrophoresis takes from several to over a dozen hours. For long-term separations, the temperature of the system is kept below 15°C. Electrophoresis may be carried out using gels of various sizes (e.g., 7.4 × 6.8 × 0.1 cm or 18.3 × 19.3 × 0.1 cm). Thickness of gels oscillates mainly between 0.1 and 0.15 cm. Usually, while optimizing separation conditions and in order to save valuable samples and reagents, it is advisable to use minisystems, whereas further analytical research is performed in a large system. The latter are characterized by a higher resolution and protein load, which is very important from the point of view of following MS analysis. Electrophoresis should be interrupted when a bromophenol blue dye added to the sample solution prior to rehydration step reaches the end of a gel.

Large-scale proteome analysis mainly utilizes vertical systems, where multiple gels are run simultaneously under the same conditions. Parallel runs and comparison of the resulting gels minimize mistakes in image analysis.

During SDS-PAGE, proteins previously resolved in an IPG strip (placed horizontally on the surface of SDS-polyacrylamide gel) enter an SDS-polyacrylamide gel once voltage is applied. Migration and separation of proteins (based on their molecular mass) occurs in a direction perpendicular to IEF and lasts as long as an electric field is applied.

Second-dimension separation is based on four main steps:

1. Preparation of a polyacrylamide gel
2. Equilibration of proteins focused on an IPG strip
3. Application of an IPG strip to a SDS-polyacrylamide gel
4. Electrophoresis

Gels should be prepared at least a few hours prior to electrophoresis. Resolution of separations is closely connected with the complexity of a sample, gel size, acrylamide/cross-linker (bisacrylamide) ratio, the percentage of acrylamide/cross-linker used to form stacking and separation gels, and the pH of the buffers used. Gradient gels, in which acrylamide concentration increases from top to bottom of the gel, resolve both large and small proteins on a single gel. Thus, they are to be chosen when complex protein mixtures are analyzed. Pore size, decreasing along the gradient, makes it possible to achieve sharp protein spots. Proteins with masses exceeding 100 kDa are typically electrophoresed in 7 to 10% polyacrylamide gels, whereas smaller molecules require 10 to 17.5% polyacrylamide. Homogeneous gels of single-percentage acrylamide are a matter of choice for separation of proteins in a narrow molecular mass range. Experimentally determined acrylamide content allows for resolving desired proteins with superior resolution. Use of the Laemmli system, employing glycine in the running buffers, enables the separation of proteins with molecular masses from 3 to 200 kDa. Furthermore, modification of the tricine gel system of Schagger and von Jagow extends that limit down to 500 Da.

STAINING AND DETECTION OF PROTEIN TRACES

Once electrophoresis has been completed and gels are fixed, separated proteins must be visualized. The fixing step, employing 50% v/v methanol in water with 10% v/v acetic acid, removes SDS and reduces background. It causes a gel to shrink, whereas staining returns it to its original size.

Staining of proteins may be carried out using Coomassie Brilliant Blue (CBB), silver, or fluorescent dye SYPRO Ruby. The fixing step allows us to distinguish proteins from a polyacrylamide gel background, due to differences in their chemical reactivity. In view of the great diversity in protein concentrations, spanning about six orders of magnitude, the step, rapid and simple in appearance, constitutes a real challenge. Alternatively, proteins may be transferred to the PVDF membranes and detected using specific antibodies (blotting; see Chapter 6).

> **Properties that an ideal stain would include** high sensitivity to the detection of even the least abundant proteins; a broad dynamic range for quantitative accuracy, reproducibility, and compatibility with techniques employed for further processing (e.g., mass spectrometry).

Unfortunately, no single stain would fulfill all the requirements.

Visualized protein profiles resulting from the same sample but stained with another dye may differ significantly, due to the fact that each dye reacts with protein in a distinct way. Often, simultaneous staining procedures are carried out on a single gel to obtain a more accurate image of proteins in a sample (e.g., a gel stained with CBB or a fluorescent dye may be further exposed to silver, or several fluorescent dyes may be used). Proteins may also be labeled prior to electrophoretic separation using radioactive labeling or a DIGE technique (Chapter 3.1). Once proteins in a sample have been visualized using the method selected, the resulting profile may be compared with images of samples of various origins presented in one of numerous databases

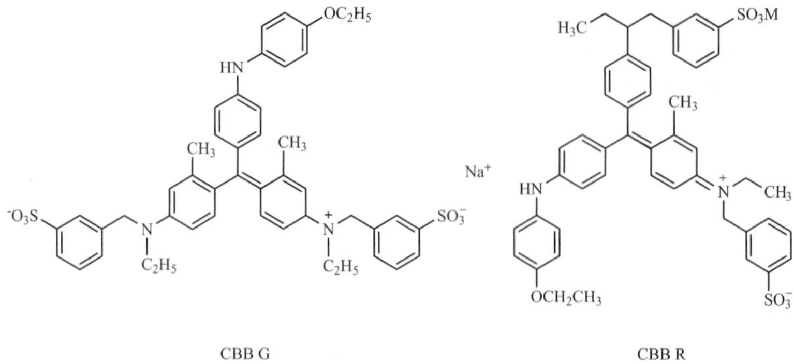

CBB G   CBB R

FIGURE 5   Chemical structure of Coomassie Brilliant Blue, forms G and R.

on the Internet (e.g., http://www.expasy.org/ch2d/ or http://www.lecb.ncifcrf.gov/flicker/).

CBB, an ionic dye, binds to proteins stoichiometrically and is thus preferred when relative amounts of protein are to be determined by densitometry. The staining technique is rapid, easy, compatible with mass spectrometry, and relatively cheap, which contributes to its popularity. In brief: A gel is soaked in a dye, usually for 30 minutes to several hours, and then destained by rinsing with water. The dye penetrates the gels slightly, creating a faint background.

CBB binds to aromatic and basic amino acids, primarily tyrosine, arginine, lysine, and histidine. The stain visualizes about 10 to 100 ng of protein per spot.

Although its sensitivity is minor compared to other methods, CBB allows displaying the broadest spectrum of proteins. Coomassie also binds to other molecules (e.g., polysaccharides), which may render the detailed analysis troublesome. Two forms of Coomassie, G and R, denote red and green hues, respectively (Figure 5). CBB G-250 provides better contrast between stained proteins and background.

Silver staining is the most sensitive nonradioactive method for protein visualization. Silver ions bind to sulfhydryl and carboxyl groups and allow detection of 0.1 ng of protein per spot. The staining itself constitutes a complex multistep process that involves carefully timed steps and requires high-purity reagents to obtain reproducible, high-quality results. The subjective endpoint of staining renders the technique far from being stoichiometric. Also, its reproducibility cannot be compared to that achieved with CBB. At present, over 100 protocols exist for visualizing proteins with the use of silver. The technique has decreased in popularity due primarily to complications with:

- *Background:* increased background signal results from the presence of DNA and liposaccharides that are also stained.

- *Quantification:* dynamic range narrower than for CBB; the stain provides a linear response with over a 10 to 40-fold range in protein concentration.
- *Mass spectrometry applications:* silver stains that do not exploit aldehyde-based fixatives render MS analysis possible, but with significantly poorer recovery of the peptides than for CBB.

Staining with zinc and copper constitute examples of reverse staining. In this technique, it is the background that is being stained; positions on the gel occupied by proteins remain transparent. The technique works due to the fact that protein-bound metal cations are less reactive than free salt in a gel and do not precipitate as fast. The technique is simple, rapid, and MS compatible, with sensitivity allowing detection of 20 to 50 ng of protein per spot. The dyes employing zinc or imidazole-zinc are considered most sensitive. However, these possess a restricted linear dynamic range, disqualifying the technique for quantitative purposes.

Autoradiography and fluorography constitute the most sensitive detection techniques, revealing 200 fg of protein per spot. The techniques rely on protein detection with the use of radioactive isotopes (e.g., ^{15}S, ^{14}C, ^{3}H, or ^{32}P) or fluorescent compounds. Radioisotopes are incorporated into proteins in vivo or in vitro. Protein derivatization with fluorophores occurs prior to IEF. For this purpose, commercially available dyes such as monobromobimane- or cyanine-based compounds are exploited (Chapter 3.1). Postelectrophoretic techniques employ mainly SYPRO Ruby staining. SYPRO Ruby, a ruthenium-based metal chelate dye, exhibits a broader dynamic range than silver while maintaining comparable sensitivity. It allows simple, background-free in-gel protein staining. Unlike silver, SYPRO Ruby is not an endpoint dye, and gel staining may be performed overnight without overdeveloping. The stain is to be visualized with advanced instruments such as a fluorescent scanner or ultraviolet light boxes, which render the technique very expensive.

An image is produced by exposing an x-ray film or emulsion to radiation of radiolabeled proteins separated on a gel. Exposure times may, however, take up to several weeks, depending on the sensitivity required. The technique also suffers from a limited dynamic range ($<10^3$). Nowadays, electronic methods for spot detection, such as phosphorimaging based on storage phosphor screens, are employed to replace x-ray film autoradiography. Autoradiography allows detecting very low levels of radioactivity in a shorter time and with a high linear dynamic range. The major disadvantage is the cost of the technique and restrictions due to working with radioactivity.

QUANTIFICATION SOFTWARE

Commercially available analysis software exploits protein pattern images acquired with the use of scanners, laser densitometers, charge-coupled device (CCD) cameras or fluorescent and phosphor imagers (depending on the protein visualization technique). Software allows comparison of digital images of the gels, spot detection, matching, and image analysis. The most widely recognized include ImageMaster, Melanie III, PDQuest, and Proteomweaver, among others. Proteins usually do not migrate to exactly the same point on each IEF or SDS-PAGE gel. Therefore, the use of complex algorithms for matching spots across the gels (warping) provides eventual shift corrections. In terms of quantification, statistical analysis may be carried out: determination of protein expression variations across the gels or between different gel populations, along with evaluation of which differences are significant. The software accurately marks proteins for gel alignment and identifies, for example, up- or down-regulation based on a precise three-dimensional analysis of protein staining intensity. Automated spot detection and matching algorithms facilitate extraction of statistically valid differences between groups analyzed. However, in many cases, gel images are of average quality and are not superimposable Therefore, user intervention is necessary to correct false matches and frequent software omissions. Image analysis ensures filtering out background noise and creating virtual gels using ideal Gaussian representations of experimental gels. The systems easily manage multiple image analysis and handle large volumes of two-dimensional gel data. The use of filtering, querying, reporting, statistical, and graphing options facilitates comparison and analysis of the results.

To confirm the results obtained with quantification software and mass spectrometry, immunoblotting or Western blotting is often carried out (Chapter 6). The technique effectively complements SDS-PAGE to visualize antibodies directed against specific proteins of interest. This enables comparative quantitative analysis of one or several proteins at a time. However, the method requires selective antibodies that might not be easily available.

Acknowledgments

This work was supported by International Centre for Genetic Engineering and Biotechnology (ICGEB) grant CRP/POL05-02.

SELF-STUDY QUESTIONS

1. Two proteins are given:
   A. A 30-kDa protein composed of a single polypeptide chain. The protein pI is 7.5.
   B. A 110-kDa protein composed of three subunits, two of 30 kDa and one of 50 kDa. The protein pI is 5.5.
      What results will be obtained if both proteins are separated via:
      a. Isoelectric focusing?
      b. SDS-PAGE?
      c. Native electrophoresis?
      For each case, draw a scheme reflecting protein distribution upon separation.
2. Distinguish the terms *isoelectric point* and *point of zero charge*.

3. Draw leucine in the ionization state that predominates at pH 8.

4. It has been shown that compared to control samples, morphine administration in rats causes variations in protein phosphorylation. Design an experiment that would allow detecting markers evoked by exposure to the drug.

RECOMMENDED READING

Berkelman T., Stenstedt T., eds. *2-D Electrophoresis: Principles and Methods.* Amersham Biosciences AB, Uppsala, Sweden, 2002.

Bierczynska-Krzysik A. Proteomic identification of brain proteins including markers evoked by exposure to morphine, highly insoluble-, and transmembrane proteins. Ph.D. dissertation. Jagiellonian University, Krakow, Poland, 2006.

BioRad. *2-D Electrophoresis for Proteomics: A Methods and Product Manual.* BioRed Laboratories, Hercules, CA.

Gorg A., *et al.* The current state of two-dimensional electrophoresis with immobilized pH gradients. *Electrophoresis* 21 (6) (2000) 1037–1053.

Gorg A., Weiss W., Dunn M.J. Current two-dimensional electrophoresis technology for proteomics. *Proteomics* 4 (12) (2004) 3665–3685.

Lopez M.F., Berggren K., Chernokalskaya E., Lazarev A., Robinson M., Patton W.F. A comparison of silver stain and SYPRO Ruby protein gel stain with respect to protein detection in two-dimensional gels and identification by peptide mass profiling. *Electrophoresis* 21 (17) (2000) 3673–3683.

Lubec G., Krapfenbauer K., Fountoulakis M. Proteomics in brain research: potentials and limitations. *Prog. Neurobiol* 69 (3) (2003) 193–211.

Oh-Ishi M., Satoh M., Maeda T. Preparative two-dimensional gel electrophoresis with agarose gels in the first dimension for high molecular mass proteins. *Electrophoresis* 21 (9) (2000) 1653–1669.

Paulson L., Persson R., Karlsson G., *et al.* Proteomics and peptidomics in neuroscience: experience of capabilities and limitations in a neurochemical laboratory. *J. Mass Spectrom.* 40 (2005) 202–213.

Schagger H., von Jagow G. Tricine–sodium dodecyl sulfate–polyacrylamide gel electrophoresis for the separation of proteins in the range from 1 to 100 kDa. *Anal. Biochem.* 166 (1987) 368–379.

Stryer L. *Biochemistry*, 4th ed. W.H. Freeman, New York, 1996.

Walker J.M., ed. *The Protein Protocols Handbook.* Humana Press, New York, 2002.

5

CAPILLARY ELECTROPHORESIS

Marek Noga

Capillary electrophoresis (CE) is a supplementary separation technique used in proteomics. Although it is an electrophoretic technique, its applications are more similar to those of liquid chromatography than to other electrophoretic techniques. As it offers some unique separation properties, CE is definitely a feasible technique for use in proteomics, most often coupled to a mass spectrometer.

Similarly to gel electrophoresis, this technique employs the motion of charged molecules in a strong electrostatic field. But instead of polyacrylamide gel, a narrow capillary filled with conductive buffer is used as a separation medium. Due to the electroosmosis phenomenon and to the high voltage applied to the capillary ends, the capillary itself acts as a "pump" providing motion to the solution inside. This effect may be utilized for the separation of both positively and negatively charged compounds, just as in systems with pressure-driven pumps.

Thanks to the special features of electroosmotic flow, capillary electrophoresis has few advantages over liquid chromatography, including very high efficiency (up to 500,000 theoretical plates) and resolution, capability for analysis of picomolar quantities of analyte, very short analysis time, rarely exceeding 15 minutes, and last but not least, very simple instrumentation. Also, miniaturization of capillary electrophoresis setup is much easier; it is expected that miniaturized analytical systems (lab-on-a-chip) will be powered by the electroosmotic flow.

The term *capillary electrophoresis* actually covers several closely affirmed analytical techniques. Basic concepts of the entire family of techniques are discussed, based on the simplest and most commonly used, capillary zone electrophoresis (CZE). Later in the chapter, other, more complex techniques will be summarized, including

Proteomics: Introduction to Methods and Applications, Edited by Agnieszka Kraj and Jerzy Silberring
Copyright © 2008 John Wiley & Sons, Inc.

micellar electrokinetic chromatography (MEKC), capillary gel electrophoresis (CGE), and capillary electrochromatography (CEC).

INSTRUMENTATION

The simplest CE setup possible (Figure 1) consists of two buffer reservoirs connected by a fused-silica capillary. The internal diameter of the capillary is about 50 μm and its typical length is between 30 and 100 cm. A platinum electrode is placed in each buffer reservoir, both connected to the high-voltage power supply. As high voltage generates flow inside the capillary, a detector is necessary to monitor the migration of sample compounds. The simplest approach utilizes an ultraviolet or fluorescence detector, just as in liquid chromatography. However, because of the high voltage applied to the capillary end of an ordinary detector, flow cells cannot be used. Instead, the detector is mounted on a transparent window prepared in the capillary sheath (Figure 2). Unfortunately, such an approach provides limited sensitivity, so more sensitive detectors using laser-induced fluorescence or mass spectrometry are used. Mass spectrometry is also the only CE detector selective enough to be used for peptide and protein identification. For that reason, most use of CE in proteomics uses a mass spectrometer as a detector. Such coupling requires a special interface between the separation capillary and the ion source, which is discussed in more detail later in the chapter.

Typical voltages applied in capillary electrophoresis range from 15 to 30 kV. Such high voltages, much higher than those used in gel electrophoresis or in an electrospray ion source, pose a serious threat to human health. Despite the fact that all commercially available instruments are well protected against electric shock, special care should be taken during system operation. Instruments coupled online to

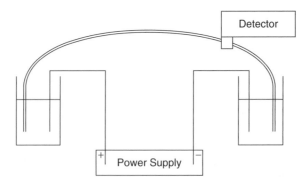

FIGURE 1   The simplest capillary electrophoresis setup, consisting of two buffer reservoirs connected by a fused-silica capillary. In both, platinum electrodes are placed where high voltage is applied. Near the capillary ends, an on-capillary UV/Vis detector is placed. Alternative detection options include the use of fluorescence and electrochemical detectors as well as a mass spectrometer whose ion source is integrated with one of capillary ends.

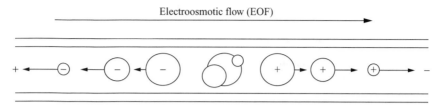

FIGURE 4   Migration order of ions in an unmodified fused-silica capillary. The fastest moving are small, highly charged cations dragged both by the electroosmotic flow and, additionally, their own electrophoresis toward the same direction. Neutral molecules move slower, dragged by the electroosmotic flow alone, and are not separated, migrating in one group. Anions, in turn, have their electrophoresis directed in the direction opposite from electroosmotic flow and reach the detector last.

Cations present in solution tend to associate near the wall, forming a *double electric layer*. The internal layer of ions (Helmholtz layer) is very strongly attracted by the negatively charged capillary wall and is therefore fixed in place. However, cations in the external parts of the layer are free to move along the capillary. After applying high voltage to both capillary ends, the cations in the external layer will start moving toward the cathode. Because all the ions are solvated, this motion will drag the buffer inside the capillary as well, powering the flow though the capillary.

This effect may be employed to "pump" the solvent though the capillary, similar to hydrodynamic flow generated by chromatography. However, there is one important difference between these two types of flow that has a significant impact on analytical process. The electroosmotic flow has a flat-plug-shaped profile, in contrast to the parabolic profile of the laminar flow generated by pressure pumps. The electrical field is uniform for each ion in a capillary, regardless of its distance to the capillary wall. However, in the case of pressure-driven flow (as in the liquid chromatography), the viscosity makes the buffer layers close to the capillary wall move more slowly than these in the center of the capillary. The main effect of the parabolic flow profile is increased mixing of subsequent solvent plugs, causing an increase in chromatographic peak width and, as a final result, a decrease in peak resolution and separation efficiency. In the case of plug-shaped flow in capillary electrophoresis, this effect is nullified. This allows the CE techniques to achieve resolution and efficiency far beyond the capabilites of the liquid chromatography.

MOTION OF MOLECULES INSIDE A CAPILLARY

Electroosmotic flow is just the pump that pulls the liquid through the capillary; it is not strong enough to separate any molecules. However, the electrical field present in the capillary directly affects the ionic species inside—through electrophoresis. Each ionic molecule has its own electrophoretic mobility and will be attracted by one of the electrodes, regardless of the electroosmotic flow. The final motion of molecules inside the capillary during the electrophoretic separation results from the

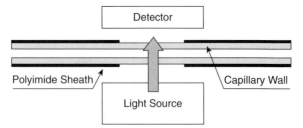

FIGURE 2 On-capillary UV–Vis detector used with CE. A layer of polyimide capi
sheath is removed to create a transparent window for direct detection of compounds migra
inside the capillary. Fused-silica capillary walls are transparent.

a mass spectrometer require even more attention, as in that case one of the capill
ends must be physically linked to the ion source.

ELECTROOSMOTIC FLOW

The most important concept behind the operation of capillary electrophoresis
electroosmotic flow (EOF) (Figure 3). Understanding this phenomenon is cruci
for conducting successful CE separation. Generally speaking, EOF is responsib
for motion of the solution inside the capillary in the same way that high-pressu
pumps are in HPLC. To generate EOF in the capillary, its walls have to be populate
with charged chemical groups (Figure 4). In the simplest example, the capillary zon
electrophoresis capillaries are made of fused silica, and their internal surface is rich i
silanol groups: Si–OH. When such a capillary is filled with a buffer whose pH is abov
2, these groups are dissociating, forming a negatively charged wall of Si–O$^-$ groups

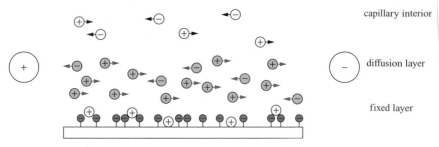

FIGURE 3 The principle of electroosmotic flow. Silanol groups (Si–OH) abundant on the
surface of the capillary wall dissociate with ejection of proton, resulting in the creation of a
strongly negatively charged surface. The cations present in solution associate near the surface,
the closest being attached strongest and immobilized near the wall. Cations situated behind
the fixed layer in the diffussion layer, however, retain the ability to move along the capillary.
After applying high voltage to the capillary ends, cations from the diffusion layer start to
move toward the cathode, dragging buffer in the capillary with them. (*See insert for color
representation of figure.*)

sum of two separate effects: electroosmosis and electrophoresis. The direction of the electrophoresis may be the same as the electroosmotic flow, resulting in increased migration velocity or the opposite of EOF, resulting in decreased migration velocity and increased migration time. The exact electrophoretic velocity depends on the molecules' charge and size.

In capillary zone electrophoresis, the fastest migrating will be the most mobile cations. They will be followed by slower cations, being more weakly attracted by the electrical field. Next, the neutral molecules will migrate, dragged only by the electroosmotic flow, and finally, followed by anions, attracted in the opposite direction by the electrical field. Of course, the most mobile anions will have the fastest electrophoretic velocity, but in the direction opposite the electroosmotic flow, thus will reach the detector last.

Although charged compounds will be separated, as they will differ either in charge or size, and the electrophoresis effect will cause them to have different total migration times, the neutral molecules will all migrate with the same velocity, equal to the velocity of the electroosmotic flow. They will not be separated and their solvent plug will only be spread by diffusion. This phenomenon explains clearly why capillary zone electrophoresis is capable of the separation of charged molecules only (see Figure 4).

The coexistence of electrophoresis and electroosmosis phenomena has one more important feature that influences the interpretation of results. It is easy to see that different electrophoretical velocities of compounds cause them also to pass the detector with different velocities. Therefore, the peak surfaces cannot be used for quantitative analysis as they are in chromatography. The fastest-migrating molecules spend the least time in the detector, and the slowest, the most. For that reason the peak surface will be normalized by division by the migration time. Only after such treatment can the peak surfaces be used to compare compound quantities.

CAPILLARY ELECTROCHROMATOGRAPHY

The inablity to separate neutral compounds in parallel with charged particles is one of the factors significantly limiting the application range of the CZE. Among the methods introduced to overcome this limitation, the most important is capillary electrochromatography (CEC). This technique is a direct hybrid of capillary electrophoresis and liquid chromatography. The separation is conducted in the capillary chromatographic column, but instead of using flow generated by high-pressure pumps, electroosmotic flow is used. Either a packed or an open tubular column with a thin layer of stationary phase applied to the surface may be used.

Such a transition of techniques has multiple advantages. High efficiency and resolution achieved thanks to the flat plug profile typical of capillary electrophoresis is combined with high selectivity of chromatographic stationary phases. In addition, because pumps do not generate the flow, the backpressure is no longer an issue. Thanks to this fact the column may be packed with smaller particles than in LC columns and/or the column length may be increased, both modifications increase the separation efficiency, resolution, and selectivity even more.

On the other hand, the use of this technique is subjected to several technical issues. First, the stability of the electroosmotic flow is problematic. The reason is that the double ion layer is generated not only on the capillary walls but also on each stationary-phase bead. The second serious issue is generation of gas bubbles on the border between packed and open parts of the capillary.

Until now the advantages of CEC over HPLC are not big enough to make the separation chemist abandon that reliable and well-established technique in favor of CEC.

COUPLING WITH MASS SPECTROMETRY

Capillary electrophoresis benefits strongly from the use of a mass spectrometer as a sensitive and selective detector. The use of MS is crucial to enabling the use of CE in certain fields of expertise, such as proteomics. The capillary electrophoresis apparatus may be coupled directly to the mass spectrometer (online) using the electrospray ion source. An off-line interface is also possible, with migrating compound being deposited subsequently on the target plate. After the separation is completed, the dried plate is placed inside the ion source of the MALDI (matrix-assisted laser desorption/ionization) mass spectrometer.

"HYPHENATED TECHNIQUES"

CE-ESI-MS (Online CE-ESI-MS)

Online coupling of capillary electrophoresis with an ESI source utilizes the fact that in both cases the electrical forces play a key role in the process. While one of the CE electrodes is placed in a buffer reservoir, the second is integrated with the ESI source. For example, if we want to conduct electrophoretic separation at 25 kV and ionization in the source is performed at 3 kV against a grounded counterelectrode, we set the voltage in the ion-source end of the capillary to 3 kV and the other end to $3 + 25 = 28$ kV.

Unfortunately, performing real analysis of this type is subject to several problems. In most cases the electrospray ionization uses the positive mode, due to its better ionization efficiency. In such cases the optimal ionization condition includes the acidic pH of the buffer. Unfortunately, low pH is very unfavorable for electrophoretic separation in unmodified fused-silica capillaries. In addition, with such narrow capillary internal diameters, intense electrochemical processes occur, causing additional, irregular pH drops. In the end, the electroosmotic flow is reduced and becomes unstable, making it difficult to conduct an analysis.

Attempts to overcome this problem caused the introduction of three different CE-ESI-MS interfaces. Each has advantages and disadvantages; thus they have slightly different applications.

1. *Sheath flow interface* (Figure 5). The key idea behind this interface is the use of additional sheath flow to overcome EOF irregularities and bolster ionization process. The end of the separation capillary is surrounded by the larger capillary, which delivers the sheath liquid. The buffers mix in the ion source. The major advantage of this setup is its ability to isolate the electrophoresis separation from the influence of the ESI source. The most important disadvantage is dilution of the separated analyte, which may hamper the sensitivity of the detection.

2. *Liquid junction interface*. In this case the system is separated into electrophoresis and mass spectrometric parts as well. The capillary end is inserted in the miniaturized buffer reservoir with an electrophoretic electrode. In the same reservoir the entrance to another capillary, guiding the liquid to the ion source, is inserted. The gap between these two capillaries is just enough to provide the electrical contact necessary for the electrophoresis process. Unfortunately, apart from technical difficulties with production, the junction causes sample dilution.

3. *Sheathless interface* (Figure 6). In this case neither sheath flow nor a junction is used, but the electrode of an electrophoresis system is literally the same as the ion source electrode. The separation capillary is tapered mechanically and its surface is covered by the conductive layer (e.g., gold, graphite dispersed in polyimide,

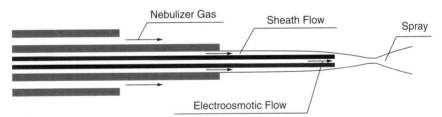

FIGURE 5 Sheath flow CE-MS interface utilizing additional nebulizer gas flow to support the ionization.

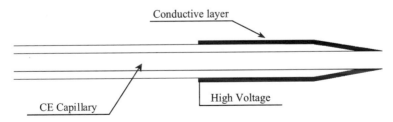

FIGURE 6 Sheathless CE-MS interface. The capillary end is tampered and covered with a conductive layer, where the high voltage is applied. The conductive layer makes it possible to create electrical contact with the electrophoretic buffer necessary both for separation in the capillary and ionization in the ESI source.

polyaniline) to provide the electrical contact for the solution. For this ion source to operate in the positive ion mode requires electroosmotic flow to be reversed and the capillary wall to be chemically modified. Unfortunately, finding proper modification of the capillary wall together with applying the conductive layer to the capillary outer surface is troublesome and limits the widespread use of this technique. On the other hand, this interface provides excellent sensitivity, as there is no sample dilution.

CE-MALDI (Off-Line CE-MS)

In the CE-MALDI case, subsequent fractions exiting the separation capillary are deposited on the target plate to be inserted into the MALDI ion source. Near the capillary end the electrophoretic buffer is being mixed with sheath flow containing the appropriate matrix. The freshly generated mixture is delivered to a piezoelectric nozzle that ejects subsequent fractions onto the selected position of the target plate. A plate prepared in this manner may be inserted in the ion source just after being dried. A standard MALDI plate may be replaced by a plate that has a hydrophobic surface with hydrophilic spots (AnchorChip). Under such conditions the buffer drop ejected by the piezoelectric nozzle is not spread on the large surface, but rather, accumulates in a small hydrophilic slot. In the end, the analysis sensitivity in improved significantly.

USE OF CE IN PROTEOMICS

Applications of capillary electrophoresis in proteomics investigations are connected primarily with peptide and protein separation utilizing a mass spectrometer as a detector. Such applications are very similar to the combination of nanoscale liquid chromatography with mass spectrometry. The CE approach, however, has some important advantages, including shorter analysis time, less sample consumption, and simpler instrumentation. For that reason, CE-MS is often used for rapid separation and identification of peptide maps.

However, the electrophoretic separation of peptides and proteins in the bare fused-silica capillaries used in CZE has some disadvantages, as such molecules interact with the charged capillary walls. This effect is most profound for protein molecules containing multiple charged residues on their surfaces. For that reason it is desirable to modify the capillary walls to reduce interactions between the capillary walls and separated molecules. However, such modifications cannot result in an uncharged surface, as that would eliminate the electroosmotic flow.

The short analysis time required for CE separations make this technique very feasible in high-thoughput applications (e.g., during clinical analyses of samples obtained from multiple patients). In multiplexed CE arrays, nearly 100 capillaries are used to separate multiple samples at the same time.

Acknowledgments

This work was supported by International Centre for Genetic Engineering and Biotechnology (ICGEB) grant CRP/POL05-02. Marek Noga was supported by a research grant N204 136 32/3396 from the Polish Ministry of Science and Higher Education.

RECOMMENDED READING

Babu S.C.V. Capillary electrophoresis at the omics level: towards systems biology. *Electrophoresis* 27 (2006) 97–110.

Harris D.C. *Quantitative Chemical Analysis*. W.H. Freeman, New York, 2002.

Niessen W.M.A. Capillary electrophoresis mass spectrometry. In *Liquid Chromatography–Mass Spectrometry*, 2nd Ed., Marcel Dekker, New York, 1998.

Simpson D.C. Combining capillary electrophoresis with mass spectrometry for applications in proteomics. *Electrophoresis* 26 (2005) 1291–1305.

Wetterhall M. Capillary electrophoresis coupled to mass spectrometry for peptide and protein analysis. In Silberring J., Ekman R., eds., *Mass Spectrometry and Hyphenated Techniques in Neuropeptide Research*. Wiley, Hoboken, NJ, 2002.

6

PROTEIN BLOTTING

Iʙᴇᴛʜ Gᴜᴇᴠᴀʀᴀ-Lᴏʀᴀ ᴀɴᴅ Aɴᴅʀᴢᴇᴊ Kᴏᴢɪᴋ

Protein blotting is an integral part of proteomics strategy and serves, among other applications, for quantitative verification of data obtained from gel electrophoresis or LC-MS runs. Since the introduction of Western blotting in the early 1980s, this and related techniques of protein blotting have become a routine extension of protein gel electrophoresis. These techniques are based on the transfer of electrophoretically separated proteins onto a membrane to which they are tightly, and usually irreversibly, adsorbed. The membrane replica ("blot") of the gel obtained can be subjected to a number of protein visualization procedures which are incompatible with the physical and chemical nature of the electrophoretic gel.

The entire protein procedure involves six steps, starting with protein separation by a selected electrophoretic technique (Figure 1, step 1). After the electrophoresis, the proteins separated are transferred onto a membrane (step 2). At this stage the protein bands can be visualized by staining with organic dyes, which, if irreversible, ends up being the simplest version of protein blotting. Usually, however, the procedure is continued toward the specific protein immunodetection (Western blotting). The membrane is blocked with neutral proteins to prevent any nonspecific binding of antibodies to the free membrane surface (step 3) and then is treated with a specific primary antibody against the proteins of interest (step 4) followed by the application of a suitable secondary antibody (step 5). The latter should bear a label that enables sensitive protein detection (step 6). In a procedural variant called *ligand blotting*, the proteins blotted onto the membrane are detected with a ligand to which they specifically bind (Figure 1).

Proteomics, a fast-developing technology of global protein analysis, instantly accommodated all major versions of protein blotting techniques. Specifically, the

Proteomics: Introduction to Methods and Applications, Edited by Agnieszka Kraj and Jerzy Silberring
Copyright © 2008 John Wiley & Sons, Inc.

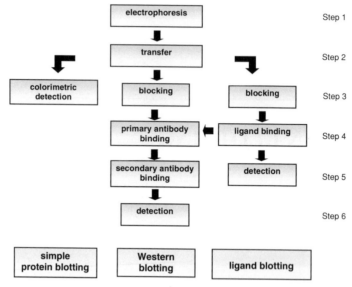

FIGURE 1   Scheme of basic protein blotting procedures.

peptides and proteins separated by routine two-dimensional polyacrylamide gel electrophoresis (PAGE) are often blotted onto membranes for specific identification,
further amino acid sequencing, or quantification of selected protein sets.

Protein blotting, particularly Western blotting, provides several advantages:

- Small volumes of reagent can be used for further detection.
- The protein diffusion on membrane is limited compared to that in electrophoretic
gel.
- The membrane makes protein detection easier, partially because it is more robust
than gel.

PROTEIN TRANSFER TO A MEMBRANE

Equipment

There are two basic systems of protein transfer to membranes. The simpler one involves free transfer due to adsorptive capillary forces. The buffer ascends through
the capillaries within cellulose paper sheets placed between a gel and a membrane
in a horizontal layout, forcing the proteins to migrate to the membrane. This type of
blotting is inexpensive and suitable for uncharged proteins, but it is time consuming
and has low transfer efficiency. An alternative method, used more frequently, is the
electrophoretic transfer of proteins from polyacrylamide gel to membrane. There
are two basic types of electrotransfer. The first requires a tank buffer apparatus
(Figure 2), and the second, a semidry blotting apparatus. In both methods a

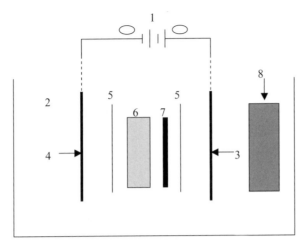

FIGURE 2   Scheme of tank system apparatus for protein transfer: 1, power source; 2, tank with buffer; 3, anode; 4, cathode; 5, adsorbent paper; 6, gel; 7, membrane; 8, cooling system.

"sandwich", consisting of adsorbent paper, gel, membrane, and adsorbent paper layers, is placed between electrodes. In the tank method, this sandwich is totally submerged in the buffer, whereas in the semidry system, buffer is used to wet the adsorbent papers, which are placed between graphite electrodes (together with the gel and membrane). The latter method (often used for minigels) is cheaper because less buffer reagent is required, but the transfer is less effective than in the tank system. For native proteins, the tank system with suitable cooling is preferred.

Membranes

A variety of membranes are suitable for protein blotting. Nitrocellulose, poly(vinylidene difluoride) (PVDF), and nylon membranes are in most common use. Hydrophobic interactions are predominantly involved in protein adsorption to the membrane, but PVDF and nylon membranes also engage electrostatic interactions. Other membrane materials, such as diazobenzyloxymethyl (DBM) and diazophenylthioether (DPT), to which the proteins bind covalently are sometimes required to minimize protein loss. Membranes for protein blotting are characterized by pore size and binding capacity. The pore size in membranes varies from 0.1 to 0.45 μm. The binding capacity is usually expressed as the amount of albumin bound to 1 cm^2 of the membrane. The binding capacities of nitrocellulose and PVDF are comparable (up to 100 or 300 $\mu g/cm^2$, respectively), but PVDF is mechanically more resistant than nitrocellulose. Positively charged nylon membranes have the highest binding capacity (up to 450 $\mu g/cm^2$) but usually give the highest background upon protein detection.

The choice of membrane is also determined by the method selected for detection and/or by the subsequent analysis to be performed. Total protein staining can be performed directly on nitrocellulose and PVDF membranes with dyes such as Ponceau S, but for the more sensitive colorimetric detection with Coomassie Brilliant Blue (CBB), the best choice is PVDF membrane. A colorimetric detection directly on nylon membrane is not possible because most general protein stains are anionic dyes and can bind to the membrane, increasing the background. PVDF membranes are suitable for Edman sequencing, amino acid analysis, on-membrane proteolysis for mass spectrometry, or for direct MALDI-TOF analysis, whereas nitrocellulose is used primarily for immunodetection.

Buffers

The choice of buffer system is determined by the protein charges, but on the other hand, it may be limited by the method of final detection. Tris-glycine is the standard buffer for protein transfer in both the tank and semidry systems. Standard buffer systems for electrophoretic transfer are listed in Table 1. In special situations, as when blotting for N-terminal sequencing of transferred protein, 10 mM CAPS buffer [3-(cyclohexylamino)-1-propanesulfonic acid] is required. Buffer pH should be above the pI value of the majority of proteins. Some proteins require supplementation of transfer buffer with additives such as methanol or sodium dodecyl sulfate (SDS).

Transfer Efficiency

The transfer efficiency depends on several factors, including the nature of the protein (e.g., size, charge), the type of membrane, the current used, the transfer time, and the

TABLE 1   Standard Buffer Systems in Protein Electrotransfer

Electrophoresis Type	Transfer Type	Buffer System	Anodic Buffer	Cathodic Buffer
Two-dimensional PAGE	Tank	Continuous	25 mM Tris, pH 8.3; 192 mM glycine; 10% methanol	25 mM Tris, pH 8.3, 192 mM glycine, 10% methanol
SDS-PAGE	Semidry	Discontinuous	I: 300 mM Tris, pH 10.4, 10% methanol II: 25 mM Tris, pH 10.4, 10% methanol	4 mM 6-aminohexanoic acid, pH 7.6, 10% methanol
IEF	Semidry	Discontinuous	I: 300 mM Tris, pH 10.4, 10% methanol II: 100 mM Tris, pH 10.4, 10% methanol	100 mM arginine, pH 7.6, 0.01% SDS, 10% methanol

buffer system (i.e., ionic strength, pH, methanol or detergent additives). The choice of membrane should be in accord with the size and charge of the protein. Commonly, nitrocellulose membranes with pore diameters of up to 0.45 μm are suitable for protein blotting (for smaller proteins, up to 20 kDa, 0.2-μm-pore membranes are recommended). However, first of all, the membrane-binding capacity should be taken into account, because very small proteins (< 20 kDa) can diffuse through some membranes but can be retained when another, more appropriate membrane is used. For membrane selection, the protein nature must also be considered: for example, some proteins, such as haptoglobin or α-glycoprotein, bind better to nylon than to nitrocellulose or PVDF membranes. Subsequent analysis after the transfer is also important. For amino acid sequencing of peptides transferred, PDVF membranes are most useful.

The composition of the transfer buffer is also essential and should be selected in accordance with the type of membrane, the nature of the proteins, and the type of electrophoresis applied for protein separation. Several additives can be introduced for transfer optimization. For example, methanol increases the binding of peptides and proteins to nitrocellulose membranes, especially when small oligopeptides are transferred. The hydrophobic interactions between proteins and the membrane are enhanced by this additive. In addition, methanol prevents protein diffusion in the gel and facilitates the SDS–protein complex dissociation. SDS (usually at concentrations of up to 0.1%) is often used in buffer systems to increase the transfer efficiency. SDS increases the relative current but also causes excessive heating, so the system needs additional cooling. At temperatures below 10°C, protein–SDS complexes may precipitate. SDS makes the elution of big proteins from the gel easier, but the binding to membrane may decrease. SDS is required when the proteins transferred are uncharged. Therefore, in many situations, a discontinuous buffer system with different anodic and cathode buffer ions is needed rather than a continuous system. These types of buffer system are applicable only in semidry transfer.

In the tank system the transfer duration is commonly 1 hour at a voltage of 10 to 20 V per centimeter of the distance between electrodes. For the semidry system, the current used can vary between 0.8 and 2.5 mA/cm^2 for 45 to 60 minutes, depending on the type of electrophoresis. The power conditions can vary depending on the buffer used and the transfer time required. The transfer time may be longer: for example, may take overnight, and then the voltage or current may be lowered. However, such a long transfer makes possible some diffusion of proteins from the membrane. When a short transfer is performed with a high field, a heating effect can denature the proteins so that cooling equipment is essential.

DETECTION

Colorimetric Detection

Nonspecific colorimetric detection can be performed, for example, when the transfer efficiency is to be checked or when a signal from a single protein must be compared against protein markers. If the protein selected is finally analyzed by immunodetection, a reversible colorimetric detection is preferred. For this purpose, Ponceau S is the most useful dye for the majority of membranes. Washing with water rinses off the dye completely, and further immunodetection can be carried out. Other types of detection, such as staining with CBB, amido black, copper phtalocyanine tetrasulfonic acid (CPTS), india ink, colloidal gold, or silver can also be carried out. All these dyes except CPTS bind irreversibly to proteins. CBB is recommended for N-terminal sequencing and amino acid analysis on PVDF membranes. The colorimetric detection is usually not suitable for a quantitative analysis because of a very variable affinity of proteins to the same dye.

Blocking Unoccupied Binding Sites

In Western blotting techniques, it is important to block the free sites on the membrane to reduce the amount of nonspecific binding of proteins during subsequent steps of the assay. At the same time, a blocking agent may improve the sensitivity of the assay by reducing the background. Different agents are used to block these sites on membranes. The most popular blocking agent is bovine serum albumin (BSA) in the concentration range 0.5 to 3%, but other protein samples, such as low-fat dried milk or serum, diluted with Tris-buffered saline (TBS) or phosphate-buffered saline (PBS), can also be used. Individual blocking buffers are not compatible with every system. The proper choice of buffer for a given blot depends on the protein transferred and on the type of conjugated enzyme used. For example, when an alkaline phosphatase (AP) conjugate is used, a blocking agent should be diluted in TBS but not in PBS, which interferes with alkaline phosphatase assay. The ideal blocking buffer will bind to the entire free surface of the membrane, eliminating background without altering the bound protein. An empirical test for the blocking step is essential for the proper choice of blocking agent.

Washing

The optimal detection in the blotting technique will be obtained when a minimal background is present with an adequate signal for the proteins analyzed. During the assay, several wash steps are necessary to remove unbound reagents for background decrease and protein signal increase. Insufficient rinsing will increase the background, whereas excessive washing may result in decreased sensitivity caused by elution of the antibody and/or antigen from the blot. Occasionally, washing is performed in a physiological buffer such as TBS or PBS without additives. More commonly, nonionic detergents such as Tween 20, Triton X-100, or Nonidet P-40 (at concentrations in the range 0.05 to 0.5%) are added to the buffer to help the removal of nonspecifically bound material.

Immunodetection

Qualitative and quantitative analysis for a specific protein transferred to membrane are usually possible by immunodetection. After protein transfer and blocking of unoccupied binding sites, the membrane is incubated with antibodies (Figure 3). First,

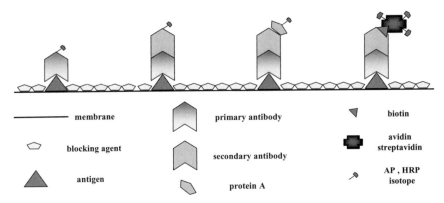

membrane

blocking agent

antigen

primary antibody

secondary antibody

protein A

biotin

avidin
streptavidin

AP , HRP
isotope

FIGURE 3   Selected setups of protein immunodetection on membrane. (*See insert for color representation of figure.*)

a primary antibody is used against the protein being investigated, and then a secondary antibody against the primary antibody is employed for further detection. The secondary antibody is commonly labeled with a radioisotope, biotin, fluorophores, or enzymes. Sometimes a secondary antibody is omitted and the primary antibody is labeled. However, this procedure is expensive, due to the large amount of labeled specific antibody required. Moreover, the label technique for antibody must be carried out on-site, because these antibodies are usually not available commercially.

Choice of the primary antibody depends on the protein to be detected. Both polyclonal and monoclonal antibodies work well for Western blotting. Polyclonal antibodies are less expensive, and they often have a high affinity for the antigen. Monoclonal antibodies are valuable, due to their specificity and purity, which result in a lower background. Crude antibody preparations such as serum or ascitic fluid are sometimes useful for blotted proteins, but the impurities present may increase the background.

Selection of the secondary antibody depends on the species of animal in which the primary antibody was raised (the host species). The host species of the secondary antibody frequently will not affect the detection. However, if the secondary antibody causes a high background in a particular assay, another host species may be chosen. Antibodies for Western blotting are typically used as dilute solutions, with the dilution range between 1 : 100 and 1 : 500,000, starting from a 1 mg/mL stock solution. The optimal dilution of a given antibody with a particular detection system must be determined experimentally.

Antibody dilutions are typically made in a wash buffer containing a blocking agent. The presence of a small amount of blocking agent and detergent in the antibody solution helps to minimize background. Due to a possible loss of proteins from the membrane during blocking and antibody treatment, the incubation time should not be longer than necessary: for example, overnight at 4°C for blocking, 1 to 2 hours at room temperature for the primary antibody, and 1 hour at room temperature for the secondary antibody.

Several radioisotopes, most frequently ^{125}I, ^{14}C, ^{35}S, and ^{33}P, can be used as labels in immunodetection. They provide excellent sensitivity, but they are expensive, have a short shelf life, and require special handling procedures. After incubation with a radioisotope-labeled antibody, the membrane is analyzed by autoradiography on radiographic film or with a phosphoimager. The use of fluorophores as labels for immunodetection is widespread. This option requires fewer steps, and the membrane can be analyzed with a phosphoimager.

An alternative label for antibodies is biotin, which can be conjugated to antibody either directly or indirectly through protein A. Biotin binds to specific proteins, avidin or streptavidin, which are conjugated with enzymes that can be detected with the proper substrates. Avidin and streptavidin are tetramers, so their use amplifies the signal and in consequence improves the sensitivity.

Enzymatic labels are used most commonly, and although they require extra steps, they can also be extremely sensitive. Two enzymes are frequently conjugated to antibodies: AP and horseradish peroxidase (HRP). Several chromogenic, fluorogenic, and chemiluminescent substrates are available for use with these enzymes. The appropriate substrate choice depends on the enzyme label (AP or HRP), the sensitivity desired, and the form of signal or method of detection desired.

The reactions of chromogenic substrates with the appropriate enzyme give insoluble colored products that precipitate on the membrane and require no special equipment for processing or visualizing. Among these substances the most popular are NBT (nitro-blue tetrazolium chloride) and BCIP (5-bromo-4-chloro-3-indolylphosphate p-toluidine salt) or Fast Red (naphthol AS-MX phosphate + Fast Red TR Salt) for use with AP. For HRP, TMB (3,3',5,5'-tetramethyl benzidine), 4-CN (4-chloro-1-naphthol), and DAB (3,3'-diaminobenzidine tetrahydrochloride) are available.

Chemiluminescent substrates are often used, particularly when higher sensitivity is required. The chemiluminescent substrates react with enzymes with light emission and offer an excellent signal for protein detection. Use of chemiluminescence allows multiple exposures to be performed to obtain the best image, or the blot can be analyzed simultaneously with another protein detection. The best system is HRP–luminol, where an enzyme oxidizes luminol through peracid salt oxidation. Special enhancers added to the system intensify the light emitted and extend the signal duration (i.e., enhanced chemiluminescence). The latter technique is one of the most sensitive detections in Western blotting, enabling detection of femtograms of protein. The signal can be analyzed by autoradiography on radiographic film or with a phosphoimager.

EXAMPLES OF APPLICATIONS IN PROTEOMICS

Ligand Blotting

After blotting to membrane, proteins can retain some ligand-binding properties, even if the electrophoresis and/or transfer was performed under denaturing conditions (e.g., SDS, methanol). It is assumed that some protein renaturation occurs during transfer and adsorption to the membrane. The ability of a given protein to bind a ligand specifically is utilized for detection. This assay bears various names, such as ligand blotting, Western ligand blotting, and ligand overlay blotting. Following membrane blocking is incubation with ligand solution. Next, the ligand can be detected with a specific antibody, or alternatively, ligand already labeled with isotopes or fluorophores is used. This technique has found uses primarily in peptide quantitative analysis, but competitive binding studies can also be carried out. For recognition of ligand-binding oligopeptides, a lot of ligands have been used, such as heparin, hormones, toxins, growth factors, nucleic acids, lipoproteins, viruses, or cells. The ligand overlay technique allowed us to discover several membrane receptors, and some functional domains in proteins were even detected.

Glycoprotein Analysis

An important version of Western ligand blotting was developed for the specific detection of glycoproteins. For this purpose, the oligosaccharide-binding proteins, lectins (also called agglutinins), are used. These proteins possess different specificities for monosaccharide residues or even for selected combinations of monosaccharides (Table 2). The proper choice of lectin array for detection on blots sometimes allows us to obtain full information about sugar components of glycoproteins. This technique, often called *lectin overlay*, may be supplemented by in situ modification of the sugar chain by appropriate glycosylases. The lectin overlay scheme includes all steps, as for a typical blot procedure; however, instead of low-fat milk, another blocking agent should be used. Good resolutions are usually obtained with isotope- and biotin-labeled lectins.

Phosphoprotein Detection by Western Blotting

Another application of Western blotting in proteomics aims at the identification of phosphorylated proteins. This is a sensitive technique often used for the specific

TABLE 2 Lectins Used for Protein Detection on Blots

Lectin Source	Specificity
Canavalia ensiformis	Mannose
Peanuts	Galβ3GalNAc
Erythrina cristagalli	Galβ4GlcNAc
Ricinus communis	Galactose
Wheat germ	GlcNAc, Neu5Ac
Limulus polyphemus	Neu5Ac
Lentil	Mannose
Helix pomatia	GalNAc
Griffonia simplicifolia	GalNAc
Soybean	GalNAc

analysis of phosphorylation sites. Usually, when proteins are digested in gel for further analysis, the amount of peptide with phosphoamino acids is small and the probe needs to be enriched before identification. In this case, Western blotting with specific antibodies against the phosphoamino acids is preferred.

Multiple Protein Detection by Qdot Western Blotting

Recently, a new, very sensitive modification of Western blotting detection was introduced for the quantitative and simultaneous analysis of proteins. In this technique, Qdot secondary antibody conjugates are used. Qdot conjugates are fluorescent materials consisting of a *quantum dot* nanocrystal coated with antibodies. The quantum dots possess a core and a shell, often CdSe and ZnS, respectively. The shell is coated with an organic layer that provides water solubility and antibody binding. Quantum dots absorb light in a wide spectral range and emit light at specific wavelengths, depending on their geometry and size. The light emitted is of high brightness and excellent photostability. This type of immunoblotting can exceed even the ECL technique in sensitivity. Fluorescence gel imaging systems are used for final detection and analysis.

Enzymatic Activity Analysis

Analysis of active enzymes after blotting requires special techniques. Enzymes possess fragile active centers since their catalytic function requires specific conformational states for substrate binding. However, some applications of blotting technique, called *renaturative catalytic blotting*, are known, especially for biological membrane-associated enzymes. Denaturation during gel electrophoresis should be partial because renaturation will be required later. After the transfer of proteins and blocking free sites on the membrane, the proteins are renatured by incubation of blot in PBS buffer with additives such as BSA. The assay of enzyme activity is performed simultaneously. In this case, an agarose film with substrate is placed on the blot, and contact between the substrate and the enzyme is facilitated. The enzymatic product is then detected in agarose film. The blot may then be utilized for the detection of enzyme following the normal immunoblotting procedure. If necessary, the specific catalytic activity can be normalized to total immunoreactivity of the enzyme.

SELF-STUDY QUESTIONS

1. Which pair of antibodies will you use for estimation of the phosphorylation degree of insulin receptor (INS-R, a tyrosine kinase) on Western blots? Why?
    A. Polyclonal mouse antiphosphotyrosine + rabbit antimouse IgG
    B. Monoclonal rabbit antiphosphotyrosine + goat antirabbit IgG
    C. Monoclonal mouse anti-INS-R + rabbit antimouse IgG
    D. Monoclonal rabbit anti-INS-R + goat antirabbit IgG

2. Propose a scheme for INS-R detection by ligand blotting.
3. Suggest a complete scheme(s) for the detection of human fibroblast growth factor receptor (FGFR, a tyrosine kinase) by Western blotting. You have available:

Blocking agent	Monoclonal rabbit antiphosphotyrosine
Wash buffer	Monoclonal rabbit antihuman FGFR
Human FGF	HRP-conjugated goat antirabbit IgG
Human FGF-HRP	HRP-conjugated rabbit antigoat IgG
Polyclonal rabbit antihuman FGF	ECL substrate for HRP

RECOMMENDED READING

Dunbar B.S. *Protein Blotting: A Practical Approach*. IRL Press, Oxford, UK, 1996.

Jackson P., Gallagher J.T. *Laboratory Guide to Glycoconjugate Analysis*. Lavoisier Editions, Paris, 1995.

Ornberg R.L., Harper T.F., Liu H. Western blot analysis with quantum dot fluorescence technology: a sensitive and quantitative method for multiplexed proteins. *Nat. Methods* 2 (2005) 79.

Pons T., Clapp A.R., Medintz I.L., Mattoussi H. Luminescent semiconductor quantum dots in biology. In: Mirkin C.A., Niemeyer C.M., eds., *Nanobiotechnology: More concepts and Applications*. Wiley-VCH, Weinheim, Germany, 2007.

Simpson R.J. *Proteins and Proteomics: A Laboratory Manual*. Cold Spring Harbor Laboratory Press, Woodbury, NY, 2003.

7

INTRODUCTION TO PROTEIN AND PEPTIDE MASS SPECTROMETRY

7.1

ELECTROSPRAY IONIZATION

PIOTR SUDER

Electrospray ionization (ESI), together with matrix-assisted laser desorption/ionization (MALDI), is an important and commonly used ionization technique in the area of biological and medical sciences. The main advantages of an ESI source are:

- Ionization of nonvolatile, high-molecular-mass compounds in the liquid phase
- Ionization under atmospheric pressure
- Soft ionization
- High sensitivity (femtomoles)

FUNDAMENTALS OF THE ELECTROSPRAY PROCESS

The breakthrough invention of ESI by M. Dole was continued by J. B. Fenn, who was awarded a Nobel prize in 2002 for outstanding applications in the biosciences. The mechanism of the electrospray process remains under investigation.

To explain ESI principles, we focus on the formation of positive ions. When a liquid passes a thin capillary to which a potential is applied, a cloud of small droplets (spray) is formed. The liquid leaving the capillary is cone-shaped (called a *Taylor cone*; Figure 1). This is the basis of the electrospray phenomenon. Positively charged ions accumulate at the surface of the droplets. The mean diameter of the droplets varies between nano- and micrometers, depending on the internal diameter of the capillary (in the nano-ESI approach, droplets are much smaller in diameter). The solvent evaporates, which leads to a decrease in distances between ions of the same

Proteomics: Introduction to Methods and Applications, Edited by Agnieszka Kraj and Jerzy Silberring
Copyright © 2008 John Wiley & Sons, Inc.

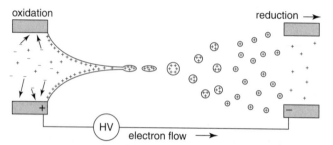

FIGURE 1   Ion formation in an ESI source. Starting from the left: delivery needle with Taylor cone, droplet formation, Coulomb fission with droplet evaporation, gas-phase ion formation in the transfer capillary. The positively charged Taylor cone is enriched with positive ions due to oxidation on the inner walls of the electroconductive delivery needle. Using a high-voltage power supply the electrons are transferred to the ion source, where reduction may occur.

charge and their accumulation at the surface of droplets. As this evaporation process continues, the density of the charges increases until the repulsion forces exceed the surface tension forces. Finally, the droplet explodes (Coulomb fission), producing smaller droplets. The process is repeated many times, until dry ions without solvent are introduced into the instrument.

Depending on the quantity of the proton (H^+)-accepting sites in the molecule, the analyte may be multiply charged (positive ions). The negative ionization mode is used when the molecule can easily release protons (e.g., when SO_3H^- groups are present). Construction of the ion source is shown in Figure 2.

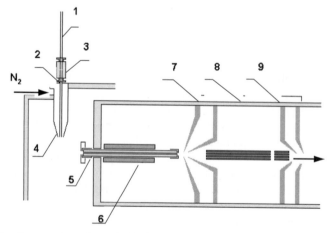

FIGURE 2   Construction of a typical ESI ion source: 1, narrow tubing delivering analyte; 2, delivery needle; 3, connector; 4, nebulizer; 5, transfer capillary; 6, heating element; 7, skimmers 1 and 2; 8, octapole 9, lenses.

TABLE 1 Differences Between ESI and Nano-ESI Sources

Parameter	ESI	Nano-ESI
Sample flow rate	1–500 (1000) μL/min	1–1000 nL/min
Flow	Supported by pump	Autonomous
Sample consumption/min	100 fmol–10 pmol	1–100 fmol
Sprayer diameter	50–200 μm	1–25 μm
Sensitivity	~10 fmol	~1 fmol (or less)
Sensitivity to salts	High	Moderate
Sheath gas	Necessary	Not used

NANOELECTROSPRAY

In a typical ESI ion source, only a small amount (less than 1%) of the introduced molecule is ionized. The main influences on this efficiency are the initial diameter of the droplets formed in the ESI needle and the flow rate. The problem with better ionization yield (hence, sensitivity) was solved by use of a nanoelectrospray needle instead of a stainless-steel device. The nanoelectrospray needle diameter is usually not larger than 20 to 25 μm, sometimes decreasing to 1 μm or even less. Adaptation of the ESI source for the purposes of nano-ESI is very simple. In most cases it is enough to remove the delivery needle with nebulizer and mount the emitter in a micro-manipulator. This device provides proper positioning of the emitter toward transfer capillary. After adjusting the geometry of both capillaries and the voltage (about 1.7 to 2.3 kV), it is possible to transform a conventional ESI source to nanoelectrospray.

Special precautions should be taken with a homemade ESI device, as high voltage may be harmful.

Nanoelectrospray requires far lower voltages than ESI because of the geometry of the system (delivery capillary needle is positioned much closer to the transfer capillary) and the volumes of samples consumed (flow rate). In a nanoelectrospray systems nebulization (i.e., spray formation) occurs due only to the voltage difference, without the use of a sheath gas. In contrast to conventional setups, there are usually few molecules in one droplet of the substance being analyzed; thus, the ionization efficiency is 10 to 100 times more effective than in typical electrospray. The two techniques are compared in Table 1.

ESI ANALYSIS IN PRACTICE

Solvents and Additives Used in ESI-Mass Spectrometry

The most popular solvents used in ESI-MS are water, alcohols, nitrils, chloride derivatives of alcanes (e.g., chloroform), and ketones (e.g., acetone). Choice of the optimal

TABLE 2   Maximum Concentrations of Salts, Buffers, and Additives in ESI and MALDI Sources[a]

Substance	Molecular Mass	Maximum Concentration in ESI (mM)	Maximum Concentration in MALDI (mM)
CHAPS	615	1.6	0.16
DTT	154	20	500
HEPES	238	1	100
NaCl	58	5	50
Na_2HPO_4	143	<1	10
NH_4HCO_3	79	50	50
SDS	288	0.35	0.35
Sodium azide	65	3	15
TFA	114	10	50
Tris	121	10	100
Triton X-100	628	<1.6	1.6
Urea	60	50	500

[a]Higher concentrations evoke ion suppression effect and signal loss.

solvent depends on two parameters: the solubility of the analyte and the ionization conditions. The common procedure is addition of a small quantity of acids (positive-ion mode), bases (negative-ion mode), or molecules that can create adducts with the sample (i.e., water, isopropyl group, etc.). The adduct calculator (with a list of possible adducts) can be found at http://fiehnlab.ucdavis.edu/staff/kind/Metabolomics/MS-Adduct-Calculator/. Formic, acetic, trifluoroacetic, and trichloroacetic acids are used for acidification. In the negative-ion mode, ammonia or sodium hydroxide is used.

As a rule, 20 to 30% methanol in water is used as a standard mobile phase. The surface tension of such a mixture is lower than that of water alone, which improves ionization efficiency significantly. There is also a general rule that volatile solvents should be used. Electrospray does not tolerate salts (especially phosphate derivatives). Some buffers may polymerize in the transfer capillary, thus clogging it irreversibly, which leads to instrument damage (see Table 2). The best solvents for protein and peptide analysis in proteomics are water solutions of alcohols and nitrils. Solvents are usually acidified. Solvent acidification to a pH value of about 4.0 using formic acid ($pK = 3.74$) causes protonation of the side groups.

ESI-Mass Spectra Interpretation

Depending on the quantity of proton-accepting (or proton-releasing) sites in the molecule being analyzed, a single or multiple charges may occur. Ionization of small molecules usually adds a single or double charge. Along with an increase in the molecular mass, the charge number increases. Thanks to this ESI feature, analysis of higher masses is possible using almost any analyzer operating at a limited m/z range (up to ca. 3000). A good example may be a protein of molecular mass 25,000 Da. This molecule can attract 20 protons to receive a charge equal to +20. An m/z values of 1251 will be observed on the spectrum:

$$\frac{m}{z} = \frac{25{,}000 + 20H^+}{20} = 1251$$

In fact, the m/z obtained must be recalculated to a real mass of the entire molecule, a process known as *spectrum deconvolution*. Note that the resolution and mass accuracy achieved by this method is much better than, for example, that obtained using MALDI-TOF (see below), especially for large intact molecules (among other factors, resolution depends on the m/z value and increases as the m/z decreases).

Here we analyze the simple case, the mass spectrum of pure, low-molecular-mass peptide (Figure 3). Such a spectrum consists of a few peaks. The charges of these peaks are usually easily readable from the mass intervals between isotopic peaks. For a singly charged molecule, the interval between isotopic peaks is equal to 1 (see the insert in Figure 3). For a doubly charged species this distance is 0.5, and for a triply charged species 0.3. A similar situation is observed for the sodium or potassium adducts, where the distance (22 or 38, respectively) is reduced by a factor of 2 or 3, respectively, depending on the number of charges (two or three) attracted by a molecule.

Mass calculation of high-molecular-mass substances from ESI spectra is more complicated. A series of peaks can be seen in Figure 4. Each peak represents the

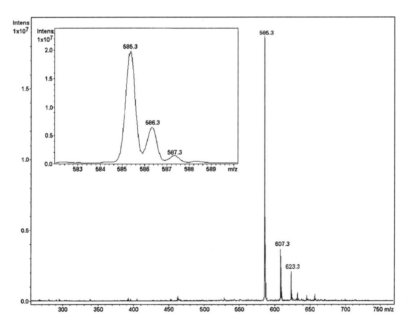

FIGURE 3   Mass spectrum of the peptide FGGFTG with a single, abundant peak at m/z 585.3 $[M+H]^+$ with sodium and potassium adducts at m/z 607.3 $[M+Na]^+$ and 623.3 $[M+K]^+$. Other peaks come from instrument noise or trace impurities in the solvents. The insert shows the isotopic distribution of the pseudomolecular ion.

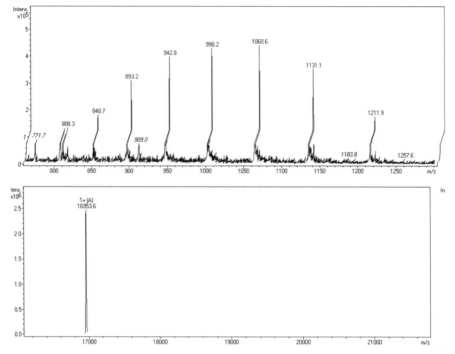

FIGURE 4   ESI mass spectrum of the horse heart myoglobin. The upper panel represents an envelope (set of peaks); and the lower panel shows the m/z ($z = +1$) value of the myoglobin after deconvolution.

same molecule, but the number of charges attracted varies. Proton attraction is a random process, and we therefore observe an envelope of ions along the spectrum (Figure 4). The number of charges depends on the number of proton acceptors, the location of those groups within the molecule (availability), the nature of the buffer, and the pH. An example of the deconvolution procedure for a spectrum obtained for horse heart myoglobin is shown in Table 3 and Figure 4.

To estimate the molecular mass, we need to calculate the charge (z) of one peak. To do this, we may use a simple equation:

$$z_n = \frac{m_{n+1} - 1.008}{m_n - m_{n+1}}$$

where m_{n+1} is the m/z of the peak at $n + 1$ charge, m_n the m/z of the peak at n charge, and, 1.008 the mass of the proton. Based on data from Table 3 and Figure 4,

$$n = \frac{1060.6 - 1.008}{1131.1 - 1060.6} = \frac{1059.6}{70.5}$$

$$= 15.03 \approx 15$$

TABLE 3 Experimental Data from ESI Analysis of 2 pmol/μL of the Horse Heart
Myoglobin

m/z	Charge Calculated	Mass Calculated	Deviation from Mean Value
771.7	+22	16,955.4	+2.4
808.3	+21	16,953.3	+0.3
848.7	+20	16,954	+1
893.2	+19	16,951.8	−1.2
942.8	+18	16,952.4	−0.6
998.2	+17	16,952.4	−0.6
1060.6	+16	16,953.6	+0.6
1131.1	+15	16,951.5	−1.5
1211.9	+14	16,952.6	−0.6
	Mean molecular mass	16,953	
	Std. dev.	1.2	

If we consider that the peak at $m/z = 1131.1$ has 15 protons attracted and its charge
is $+15$, the molecular mass of the substance may be calculated as follows:

$$M(\text{g/mol}) = m^*n - n$$

and for a charge of $+15$ and $m/z = 1131.1$, we have

$$M(\text{g/mol}) = 1131.1^*15 - 15 = 16,966.5 - 15 = 16,951.5\,\text{g/mol}$$

The result obtained from such a calculation is not precise enough. Deconvolution
should be repeated for all abundant peaks and the standard deviation should be used
(Table 3). For example, the proper result should be:

$$\text{myoglobin MW} = 16,953 \pm 1.2\,\text{Da}$$

Therefore, due to a rather laborious procedure, deconvolution software for automatic processing is usually provided by the manufacturers.

More difficult to interpret are spectra obtained from mixtures. Figure 5 presents
such a problem, where a peptide map released from one protein is shown. Such a
spectrum may contain peptides that have different charges, and the first task is to
find all ions belonging to the peptide. These peaks should be excluded from the next
searches and the procedure repeated until all ions are assigned. Interpretation may be
complicated by such phenomena as ion suppression, adduct formation, and similar
or identical m/z ratios for various components. In some cases, analysis of complex
mixtures containing a few proteins may even be impossible. Peaks belonging to
distinct proteins may overlap on the spectrum, what usually leads to false-positive
results. An automatic deconvolution procedure may take ions belonging to different
components for calculation. In such a case, application of hyphenated techniques
(e.g., HPLC-ESI-MS) is strongly recommended.

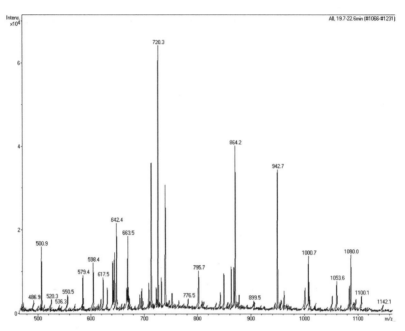

FIGURE 5   ESI mass spectrum of the peptide map of human serotransferrin after trypsin cleavage. Automatic deconvolution of such a mixture often leads to a false result, as the software may select ions that belong to various components.

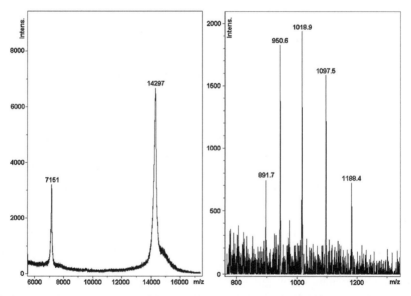

FIGURE 6   Comparison of MALDI and ESI spectra of the same molecule. Left panel: MALDI; right panel: ESI. Note the differences between signal intensities as well as background noise (compare the scales at the Y-axes). Fifty femtomoles per microliter of bovine lysozyme was introduced to both instruments. The molecular mass of the molecule is 14,248 g/mol.

It is also important to discuss the sensitivity of the ESI compared to, for example, the MALDI method. ESI is less sensitive than MALDI, due to formation of the multiply charged species. Analyzing 1 pmol of the 20-kDa molecule using a MALDI source, the resulting spectrum will contain one or two peaks. This means that the signal is distributed among one or two ions. In the ESI method, the same amount of the substance will result in formation of multiply charged species (10 to 20, depending on the mobile phase and molecular mass). As the total signal will be distributed among many ions, the sensitivity of this method must be lower (see Figure 6).

Acknowledgments

This work was supported by International Centre for Genetic Engineering and Biotechnology (ICGEB) grant CRP/POL05-02.

SELF-STUDY QUESTIONS

1. Why are ion formation and its transfer to the analyzer in the ESI source not obscured by the molecules in air? In the case of a MALDI ion source, for example, higher air pressure inside the ionization chamber affects the ionization process.
2. An ESI source works as an electrochemical cell. When it operates in a positive-ion mode, negatively charged ions participate in delivery needle oxidation. As an effect, positively charged ions travel to the transfer capillary and other elements of the instrument. But what is the fate of electrons generated during an oxidation process?
3. Why are data received after ESI-MS analysis of proteins usually more accurate than those obtained from a MALDI-TOF instrument?
4. Why do nanoelectrospray needles have a much shorter lifetime (hours) than that of typical electrospray stainless-steel delivery needles (years)? What are the major factors that influence nanoelectrospray needle life?

RECOMMENDED READING

Annesley T.M. Ion suppression in mass spectrometry. *Clin. Chem.* 49(7) (2003) 1041–1044.

Cech N.B., Enke C.G. Practical implications of some recent studies in electrospray ionization fundamentals. *Mass Spectrom. Rev.* 20(6) (2001) 362–387.

Cole R.B. Some tenets pertaining to electrospray ionization mass spectrometry. *J. Mass Spectrom. from.* 35(7) (2000) 763–772.

Jessome L.L., Volmer D.A. Ion suppression: a major concern in mass spectrometry. *LC GC North Am. Suppl. S*, June (2006) 83–89.

Johnstone R.A.W., Rose M.E. *Mass Spectrometry for Chemists and Biochemists*, 2nd ed. Cambridge University Press, New York, 1996.

Kebarle P. A brief overview of the present status of the mechanisms involved in electrospray mass spectrometry. *J. Mass Spectrom.* 35(2000) 804–817, 2000.

Smith D.R., Moy M.A., Dolan A.R., *et al.* Analytical performance characteristics of nanoelectrospray emitters as a function of conductive coating. *Analyst* 131(4) (2006) 547–555.

7.2

MATRIX-ASSISTED LASER DESORPTION/IONIZATION

Agnieszka Kraj and Marcus Macht

ESI and MALDI are the two ionization techniques that have the greatest impact on biological mass spectrometry and in the field of proteomics. MALDI is widely used not only for protein and peptide analysis but also for studies on oligonucleotides, oligosaccharides, and for large polymers and their mass distribution. Matrix-assisted laser desorption/ionization (MALDI)–mass spectrometry (MS) was developed in the 1980s by Karas and Hillenkamp and also by Tanaka as a new ionization method for the analysis of peptides and proteins. For the first time, proteins with masses higher than 10 kDa were measured as intact molecules. Similar to electrospray, MALDI causes little or no fragmentation of the compounds being analyzed and belongs to the *soft ionization techniques*. MALDI is advantageous in the analysis of complex mixtures, facilitating identification of protonated molecular ions in general.

MALDI IONIZATION PROCESS PRINCIPLE

In the MALDI ionization process, the crucial role is played by a low-molecular-mass organic matrix added in large excess to the sample. Matrix, cocrystallized with sample, will strongly absorb laser wavelength (mostly ultraviolet or infrared). Irradiation causes the matrix molecule to expand into the gas phase together with intact protein molecules. The most possible ionization mechanism is described as excited-state proton transfer between a photoionized organic matrix and a sample molecule [see equation (1)], but other possibilities (e.g., multiphoton ionization, thermal ionization, electron transfer) are still under discussion. The mechanisms of ionization in

Proteomics: Introduction to Methods and Applications, Edited by Agnieszka Kraj and Jerzy Silberring
Copyright © 2008 John Wiley & Sons, Inc.

MALDI are well described in a review article by R. Zenobi and R. Knochenmuss:

$$\begin{aligned}
&M + h\nu \rightarrow M^* \\
&M^* + A \rightarrow (M - H)^- + AH^+ \\
&M^* + M \rightarrow (M - H)^- + MH^+
\end{aligned} \tag{1}$$

where M represents a matrix and A an analyte. Matrices should promote proton transfer in the gas phase, and one of the examples of this type of compound is salicylic acid. Effective matrices are compounds with a hydroxyl group in the ortho position toward any of the carbonyl derivatives (CO, COOH, $CONH_2$).

IMPORTANCE OF THE MATRIX

The choice of matrix (Table 1) is important, depending on the analyte. A wide range of matrices is now available, but selection of the proper compound is still primarily a question of the investigator's intuition. In protein analysis the most suitable are benzoic acid and its derivatives, and cinnamic acid. Also, 2-(4-hydroxyphenylazo)benzoic acid is considered as a good matrix in peptide and protein analysis (Juhasz et al., 1993). Compounds such as picolinic acid (oligonucleotides), succinic acid (oligonucleotides), and dithranol (synthetic polymers) are other examples of matrices.

General properties of a good-quality, appropriate matrix are:

- Solubility in the same solvents as the analyte
- Stability in the vacuum environment
- Prevention of cluster formation
- Absorption of a desired laser wavelength
- Cocrystallization with sample
- Promotion of analyte ionization

Several techniques for sample and matrix deposition on the target plate have been described, the most popular being the dried-droplet method. A drop of aqueous matrix solution is mixed with sample and left to dry. This method is suitable for on-spot sample precleaning simply by washing out buffers and other nonvolatile compounds before measurement. Another, more advanced method is spraying a matrix on the target plate using an electrospray approach. This type of sample deposition generates the most homogeneous layer of crystals and is often employed in uniform matrix deposition for imaging mass spectrometry (see Chapter 7.3). Among other crystallization methods are fast evaporation, sandwich, overlayer deposition, slow crystallization, ultrathin layer, and vacuum drying. Dried MALDI samples are generally stable and can be stored, for example, in a sealed container with an inert gas or in a refrigerator (vendors are now producing single-use MALDI inserts). It is important to keep the pH of matrix solution below 4; thus, addition of 0.1% TFA is strongly recommended.

TABLE 1   Examples of Matrices in the Analysis of Various Types of Molecules

Matrix	Structure	Application
α-Cyano-4-hydroxycinnamic acid		Proteins and peptides up to 5 kDa
2,5-Dihydroxyacetophenone		Proteins 8 to 100 kDa, glycoproteins, and AnchorChip preparations
2,5-Dihydroxybenzoic acid		Peptides, polymers, phosphoproteins, glycoproteins, and carbohydrates
Sinapinic acid		Peptides, proteins, and polar polymers
Dithranol		Nonpolar polymers and fullerenes
2,4,6-Trihydroxyacetophenone, 2,3,4-trihydroxyacetophenone		Oligonucleotides smaller than 6.5 kDa[a]
3-Hydroxypicolinic acid		DNA and RNA samples below 30 kDa

(*Continued*)

TABLE 1   Examples of Matrices in the Analysis of Various Types of Molecules (*Continued*)

Matrix	Structure	Application
Succinic acid		IR-MALDI
1,5-Diaminonaphtalene		Proteins and peptides

[a]2,4,6-Trihydroxyacetophenone and 2,3,4-trihydroxyacetophenone, as a mixture of two isomers, can be used as a MALDI matrix.

MALDI ionization mostly generates singly protonated pseudomolecular ions $[M + H]^+$ (primarily by an excited-state proton transfer mechanism), but quite often the cationized molecules are visible in the form of the adduct $[M + Na]^+$ or $[M + K]^+$ (gas-phase cationization). Sodium and potassium ions are commonly present in solutions (especially in biological fluids) or come from, for example, glassware, resulting in mass spectra as cationized adducts. This observation can be utilized to our benefit. In the case of less polar compounds (e.g., polymers), it is best to use metal salts, such as silver or copper, to improve ionization efficiency. Addition of ammonium salts to the matrix can prevent matrix clustering. Some surfactants (e.g., cetrimonium bromide) are used as matrix additives to suppress matrix-related ions in mass spectra (in the case of low-molecular-mass compound analysis).

Most MALDI matrices are designed for a UV nitrogen laser emitting light at 337 nm. As an alternative, an infrared laser can be used (e.g., Er:YAG emitting at 2.94 μm). Analysis of oligonucleotides and some noncovalent compounds takes advantage of the IR laser. This type of ionization is commonly used in atmospheric pressure-MALDI. IR laser ionization does not require a specific matrix, and water can be used instead, which greatly simplifies analysis.

Solid matrices have some drawbacks. They cocrystallize with sample, yielding spots with unevenly distributed molecules. This results in significant changes in ion signal abundances and inefficient sample consumption. A suitable alternative might be a liquid matrix, providing homogeneous sample mixing, good signal abundance, and lower sample consumption. The principle of liquid matrices is based on an old concept of sample preparation for fast atom bombardment (FAB), where glycerol was commonly used. The important feature of such support is its low viscosity and low vapor pressure, making this reagent suitable for analysis in high vacuum. The laser light–absorbing chromophores (all presently available matrices, including

CHCA, SA, and DHB) are dissolved in glycerol and mixed with sample. Methanol is also added to concentrate the effect of the spot as the solvent evaporates. It is also possible to adjust the pH of such a matrix, which is vital for, for example, analysis of noncovalent complexes. Also, all other methodological "tricks" taken from the FAB concept, such as on-spot tryptic digestion, and chemical derivatization, can be performed. Liquid matrices are under constant investigation, and their features can be very promising for MALDI applications.

SAMPLE PREPARATION

In MALDI sample preparation, it is recommended that salts and contaminants be removed prior to measurement, but this ionization technique is less sensitive then ESI to salt content, so millimolar concentrations of Tris are acceptable and the spectra are not much affected. With a higher content of additives, especially detergents such as SDS, ion suppression is pronounced, causing poor spectra acquisition or no spectra at all (see also Chapter 7.1).

Target plate selection should be chosen carefully for successful measurement. In addition to the typical stainless-steel surface, gold-covered and AnchorChip [a metal probe covered with hydrophobic layer and small hydrophilic anchors (200 to 800 μm) for sample concentration] surfaces are used, often prespotted with matrix and standards. Moreover, Bruker Daltonics introduced a novel sample prefractionation for large proteins. This method utilizes magnetic beads with a surface that has special functionality, making it possible to capture desired proteins. Additionally, the total process of analysis can be automated with robotics for the most robust, highly reproducible analysis.

ANALYZERS: GENERAL REMARKS

During the last decade, fast development in the area of mass analyzers, in conjunction with MALDI mass spectrometry, has been observed. For the MALDI source, the most often used are time-of-flight (TOF) and hybrid TOF analyzers, with particular focus on Q-TOF and TOF/TOF, as these technical solutions provide MS/MS capabilities. Crucial parameters, which describe the performance of TOF analyzers, such as resolution and accuracy, exceed $R = 12,000$ in modern machines, and 1 ppm or even better (with internal calibration) in routine analysis. Additionally, unlimited mass range, speed of analysis, and sensitivity (all ions can be measured), are desirable features of this analyzer in protein and peptide measurement.

Developments in recent years show that connecting an external MALDI source with high-resolution analyzers (e.g., FT-ICR) is a robust analytical platform. An FT-ICR analyzer works routinely at a resolution of 100,000 and guarantees mass accuracy far below 1 ppm. Such an analyzer is beneficial for solving an entire isotopic distribution of a measured protein, thus providing information on monoisotopic mass (the exact mass of the most abundant isotope) as well as for the rest of the isotopic cluster.

PRINCIPLE OF THE TOF ANALYZER

The principle of the TOF analyzer can be described briefly as follows. The mass-to-charge ratio is derived from the time of flight of ions through an analyzer tube in a vacuum environment. The ions are extracted and accelerated through a system of optics and controlled by applying proper voltage. Ion optics is also the first stage of focusing ions. Ions are pulsed from the source (or first analyzer in the case of Q-TOF) and separated in a field-free region until they reach a detector. The ions are separated according to their m/z ratios, and lighter ions travel faster.

TOF analyzers can work in both linear and reflectron modes (Figure 1). In the linear mode the detector is situated parallel to the source. The linear mode is characterized by lower resolving power than in the reflectron mode. To improve the resolution of a TOF analyzer, it is possible to pulse ions further in a reflectron (set of electronic mirrors), which compensates for dispersion of ions having the same m/z values. Because proteins with masses higher than 10 kDa undergo metastable decay, the reflectron mode is applicable only for molecules not exceeding 10 kDa.

Ion extraction can be delayed to improve resolution. Delayed extraction compensates for kinetic energy dispersion by delaying ions formed in the source. This results in the production of a "package" of ions, which are then better focused.

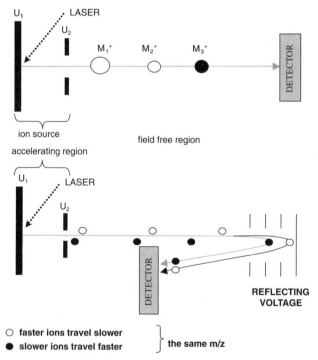

FIGURE 1 Basic scheme of the linear (upper) and reflectron (bottom) modes of a TOF analyzer.

MALDI APPLICATIONS IN PROTEOMICS

MALDI-mass spectrometry is a reliable technique in proteomics and can be used for accurate mass determination of intact proteins, a fast sample purity check, screening for posttranslational modifications or after digestion with enzyme (e.g., trypsin) for analysis of peptide maps. Additionally, MALDI can be utilized in the top-down sequencing of proteins, and analysis of noncovalent interactions. Imaging mass spectrometry is also one of the most important uses of MALDI in proteomics.

Analysis of Intact Proteins

One MALDI goals is intact protein profiling in biofluids in tissue homogenates. Often, the spectra are limited to the most abundant proteins, and to increase sensitivity for less abundant ones, extensive prefractionation steps, modification of target plate (SELDI), affinity enrichment (ClinProt), or on-target preconcentration (AnchorChip) are necessary.

An important advantage of MALDI mass determination is rapid analysis compared to other methods. To determine the molecular mass of proteins with high mass accuracy, a MALDI source can be connected with an FT-ICR analyzer, which gives more reliable information on molecular mass. An FT-ICR analyzer also offers robust fragmentation techniques [e.g., sustained off-resonance irradiation collision-induced dissociation (SORI-CID), infrared multiphoton dissociation (IRMPD), electron capture dissociation (ECD) (see Chapter 7.4)] that can be applied to MS-based aspects of proteomics (top down and bottom up). Figure 2 is an example of an intact protein analysis. There are also interesting techniques for large molecule fragmentation using MALDI-TOF instrumentation combined with in-source decay (ISD) (see the recommended reading and Chapters 7.4 and 7.6). An example of such analysis of thioredoxin (ca. 12 kDa) is shown in Figure 2. A protein sample is subjected to increasing laser power and primarily c and y ions are observed.

The sample preparation procedure requires some experience, but its complete description is outside scope of this book. It should also be noted that all ISD approaches accept pure components and that mixtures cannot be analyzed using this method.

Analysis of Peptide Maps

MALDI can be used extensively for peptide map analysis. Purified protein is digested with enzyme of known specificity, preferably trypsin. Trypsin cleavage generates a finite number of peptides with an N-terminus of NH_3^+ and Lys or Arg on the C-terminus. The average peptide carries two or three charges, and the entire map is a specific fingerprint of particular protein. For identification, sequence databases (SwissProt, MSDB) are searched (more information in Chapter 7.5).

In the peptide map approach, the challenges are (1) protein purity (for MALDI it is usually two-dimensional electrophoresis), (2) a reliable database for identification, and (3) experience in identification of posttranslational modifications that may obscure a correct match. Peptide maps are not optimal for small proteins, due to an

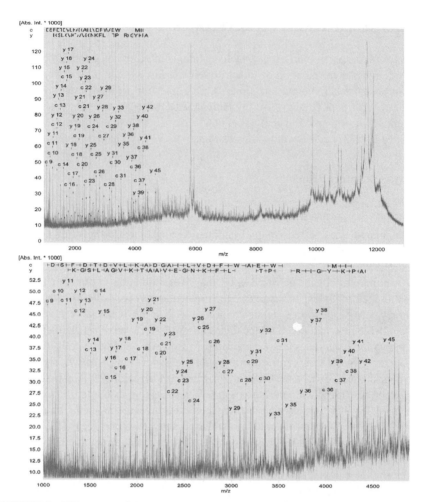

FIGURE 2   ISD spectra of the protein thioredoxin (ca. 12 kDa). (*See insert for color representation of figure.*)

insufficient number of tryptic peptides. FT-ICR as an analyzer can be a solution here, due to excellent resolution and accuracy. Figure 3 shows an example of a MALDI peptide map.

Sequencing of Intact Proteins (Top-Down Proteomics)

In the top-down approach, the intact protein is introduced and ionized in an ion source such as ESI or MALDI, and then fragmented. This provides information on the molecular mass of peptides or proteins. MALDI-MS/MS can be applied for intact protein sequencing, and the primary sequence of the protein being investigated as well as its modification can be identified (see Chapter 7.4). Fragmentation of large

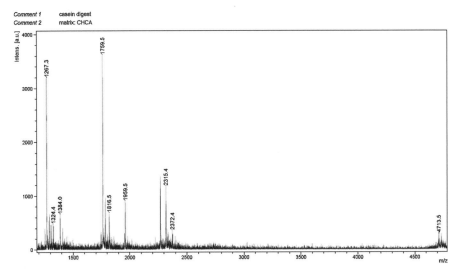

FIGURE 3   Peptide map of casein. Protein was digested with trypsin and spectrum registered in the linear mode.

proteins in the gas phase is still challenging; therefore, electron capture dissociation and electron transfer dissociation techniques are becoming more popular.

Analysis of Posttranslational Modifications

Posttranslational modifications are important for the final function of proteins. Nearly each protein undergoes modification with covalently bound group(s) (e.g., phosphorylation, acetylation, or farnesylation) (see Chapter 8 for a detailed description of posttranslational modifications). MALDI-mass spectrometry is an alternative tool to ESI for identification and localization of possible modifications. Figure 4 presents the detection of phosphopeptides already in the MS mode (full scan).

For known fragments it is possible simply to calculate the mass difference between the mass of peptide and the fragment ion observed. This approach is correct when stable, covalent modifications are investigated as, for example, acetylation indicated by a mass difference of 42 Da. Other modifications and mass differences are given in Table B1 (Appendix B) and in Chapter 8.

Imaging Mass Spectrometry

The concept of molecular scanning of tissues is one of the newest applications of the MALDI technique, not only in proteomics. The method provides information on the distribution of particular peptides and proteins in tissue sections. The advantage is that neither homogenization nor sample enrichment is needed. Moreover, this method measures the distribution of various molecules in the tissue and their colocalization.

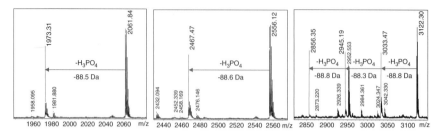

FIGURE 4   Observation of metastable neutral losses of phosphate from three different phosphopeptides. Resolution of the metastable signal is reduced significantly compared to the precursor signal, allowing for identification of a metastable fragmentation in the MS mode. The apparent mass difference between the precursor and the metastable fragment for a given neutral loss is relatively constant, as shown for the three precursor masses, and can be used to identify the nature of the neutral loss.

A detailed description of imaging mass spectrometry is given in Chapter 7.3 and in the recommended reading. The procedure has four steps:

1. A frozen tissue sample is sliced.
2. The slices are placed on a MALDI target plate.
3. A matrix is applied to the sample.
4. Mass spectra are acquired.

In this technique, data processing is a crucial factor influencing the robustness of the method, demanding sophisticated algorithms and advanced chemometrics to analyze a vast number of data.

Generally, we can distinguish a *profiling part* when only a few selected spots are analyzed and data are compared between, for example, healthy and diseased subjects, and an *imaging part*, when a whole section is scanned with defined resolution at fixed intervals. MALDI imaging can be used in peptide and protein profiling in tissues and specific regions of organs.

Acknowledgments

The kind help of Bruker Daltonics during preparation of this material is highly appreciated.

SELF-STUDY QUESTIONS

1. Why can we observe multiple peaks on a MALDI spectrum derived from a pure peptide?
2. For large peptides, why is the difference between the monoisotopic mass calculated and the highest mass observed on the spectrum 1 Da?

3. Why do we need sample preparation prior analysis in MALDI? Suggest the work flow for analysis of a crude protein extract using MALDI.

4. What type of analyzer would you choose to identify two peptides with sequences that differ by only one amino acid: the first peptide containing Q and the second containing K.

5. Which mode of the TOF analyzer (linear or reflectron) should you use for the analysis of large proteins?

6. What types of fragmentation are possible in MALDI-TOF instruments during peptide analysis?

7. Is a MALDI source linked to an FT-ICR analyzer suitable for peptide mass fingerprinting?

8. Is it possible to achieve fragmentation of peptides in an FT-ICR analyzer?

RECOMMENDED READING

Hillenkamp F., Karas M., Beavis R.C., Chait B.T. Matrix-assisted laser desorption/ionization mass spectrometry of biopolymers. *Anal. Chem.* 63(24) (1991) 1193–1196.

Johnstone R.A.W., Rose M.E. *Mass Spectrometry for Chemists and Biochemists*, 2nd ed. Cambridge University Press, New York, 1996.

Kowalski P., Stoerker J. Accelerating discoveries in the proteome and genome with MALDI-TOF-MS. *Pharmacogenomics* 1(3) (2000) 359–366.

Mann M., Hendrickson R.C., Pandey A. Analysis of proteins and proteomes by mass spectrometry. *Annu. Rev. Biochem.* 70 (2001) 437–473.

Reyzer M.L., Caprioli R.M. MALDI-MS-based imaging of small molecules and proteins in tissues. *Curr. Opin. Chem. Biol.* 11(1) (2007) 29–35.

Silberring J., Ekman R. *Mass Spectrometry and Hyphenated Techniques in Neuropeptide Research*. Wiley-Interscience, Hoboken, NJ, 2002.

Siuzdak G. *The Expanding Role of Mass Spectrometry in Biotechnology*. MCC Press, San Diego, CA, 2003.

Zenobi R., Knochenmuss R. Ion formation in MALDI mass spectrometry. *Mass Spectrom. Rev.* 17(1998) 337–366.

7.3

IMAGING MASS SPECTROMETRY

Małgorzata Iwona Szynkowska and Jerzy Silberring

Mass spectrometers are devoted to analyzing and identifying molecules. Such analysis allows for complete simultaneous identification of substances, but not their spatial and/or lateral distribution in various materials or tissues. Imaging mass spectrometry is a relatively new methodology, aiming at identification of molecules spatially distributed on the surface of a given material or tissue. The immediate benefit of this strategy might be direct scanning of profiles to reveal distribution of a particular substance directly in the tissue (e.g., after tissue biopsy) or identification of the fingerprints (e.g., for forensic purposes). This method is often referred to as a *molecular scanner*. This puts a new light on our knowledge, with potential applications in medicine, pharmacology, and forensic sciences.

Methods used presently that have a potential advantage in scanning are:

- Time-of-flight secondary ion mass spectrometry (TOF-SIMS)
- Matrix-assisted laser desorption/ionization–time-of-flight (MALDI-TOF)
- Atmospheric pressure MALDI
- Desorption by electrospray (DESI)

Major advantages and disadvantages, together with potential applications, are described below.

TIME-OF-FLIGHT SECONDARY ION MASS SPECTROMETRY

In the early 1970s the acronym SIMS (*secondary ion mass spectrometry*) was introduced, and further development of this method as TOF-SIMS took place primarily

Proteomics: Introduction to Methods and Applications, Edited by Agnieszka Kraj and Jerzy Silberring
Copyright © 2008 John Wiley & Sons, Inc.

in Alfred Benninghoven's group in the Physics Department of the University of Münster. In cooperation with many industrial and research workers, they developed TOF-SIMS into one of the most powerful techniques for surface characterization of solids, enabling determination of chemical surface composition, both elemental and molecular. In recent years, development of the TOF-SIMS/laser-SNMS (secondary neutral mass spectrometry) instrument was focused on its applications in the life sciences.

Fundamentals

SIMS is based on the information transferred by secondary ions emitted from a surface examined after its bombardment by short pulses of fast primary ions. The interaction between the primary ions [e.g., Ga^+, Cs^+, Au_n^+, Bi_n^+, Bi_n^{2+} ($n = 1$ to 7), O_2^+, Ar^+, Xe^+] and the material analyzed causes the emission of such surface particles as positive and negative secondary ions (i.e., elements, isotopes, fragments, intact molecules, clusters) and electrons (Figure 1). Ions are ionized via an attachment of protons or loss of hydrogen and small functional groups. The fragmentation process, which may occur during the analysis, gives a characteristic pattern on the secondary ion mass spectra of the sample studied.

The primary ion generates an intense but short-lived collision cascade, and many atoms of the matrix are relocated. Some of the atoms near the surface receive enough energy to leave the surface. So as a result of bombardment, the surface of the sample undergoes changes connected not only with the loss of secondary ions, but also with implantation of primary ions and atom relocation (mixing). All these changes are characterized by a *damage cross section*, which is described by the average size of the surface area modified as a consequence of single ion impact.

Masses of emitted secondary ions, elements, and molecules are determined by measurement of their flight time from the surface of the sample being investigated to

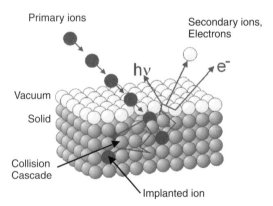

FIGURE 1   The principle of SIMS analysis. [Reprinted from Paryjczak and Szynkowska (2003) with permission.]

the detector. The time needed to reach a detector is connected directly to the value of the mass-to-charge ratio (m/z). The lighter ions arrive before the heavier ones, and a mass spectrum is recorded. A TOF analyzer allows simultaneous recording of all ions detected with high mass resolution, greater than 9000 at $m/z = 29$ and high sensitivity. The energy of primary ions ranges between 5 and 25 keV. The time of impulse of these ions lasts from 0.6 to 100 ns. The distance between impulses can be changed from 50 to 300 µs. The angle of incidence of primary ions varies from 0 to 60°. Research can be carried out in the range of temperatures from −130 to 600°C. In present versions of TOF-SIMS spectrometers, nonconducting samples (insulators) can be studied.

The Ionization Process

Ionization is relative to the chemical state of a sample surface during the sputtering process. The formation of secondary ions depends strongly on the electronic differences between small parts leaving the surface, and the surface itself. The efficiency of secondary ion formation is described by the *useful yield*, defined as the number of secondary ions generated divided by the number of corresponding parent species disappearing from the surface during primary ion bombardment. It depends on the chemical nature of surface species investigated, matrix effects, and properties of primary ions, such as mass and energy. Useful yield, sensitivity, and quantification of the method can be improved significantly by using laser beams (laser-SNMS) to ionize neutral particles sputtered from the sample surface.

Construction of a TOF-SIMS Spectrometer

The main construction elements of TOF-SIMS spectrometers are (Figure 2):

- Sources of primary ions
- System of ion optics
- Time-of-flight analyzer
- Detector

Ion sources, in general, produce an ion beam from a gas, a liquid, or a solid. The commonly available ion sources used in TOF-SIMS are:

1. Liquid metal ion sources (LMISs) (metals such as Ga, In, Au_n, Bi_n)
2. Surface ionization sources (Cs)
3. Gas sources:
   a. Electron impact (noble gas ions, oxygen, SF_5, C_{60})
   b. Plasma sources (noble gas ions or oxygen)

LMIS and surface ionization sources have their greatest use in TOF-SIMS spectrometers. A gallium gun has been used most frequently in commercial instruments

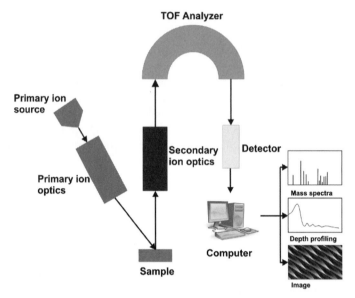

FIGURE 2   Schematic diagram of a TOF-SIMS instrument. [Reprinted from Paryjczak and Szynkowska (2003) with permission.]

as a liquid-metal ion source due to its low melting point and low reactivity with oxygen. Using different ion sources, the efficiency of various secondary ions can be affected; for example, cesium enhances the yield of electronegative ions, whereas oxygen enhances electropositive ions. Gas guns are used primarily in etching in dynamic SIMS.

Analyzers of Secondary Ions

In SIMS spectrometers, three types of mass analyzers are used: quadrupole, magnetic, and time-of-flight.

Static and Dynamic TOF-SIMS Modes

TOF-SIMS spectrometers enable studies in two main modes, static and dynamic. At the beginning, SIMS spectrometers were equipped with magnetic or quadruple mass analyzers, which made it possible to work only in the dynamic mode. During analysis of the sample, the destruction of many monolayers took place and essential chemical information was lost. So the most popular aspect of the SIMS technique was analysis of the depth profile of the inside of a sample. Introduction of the TOF analyzer opened new perspectives on using the static SIMS mode in many analytical applications as well as in fundamental research. It is a quasi nondestructive analysis which makes it possible to obtain information from the top monolayers of a surface. In this mode the primary ion dose is smaller than 10^{13} ions/cm^2, and only about 1%

of the surface being studied is destroyed. The typical dose value varies from 10^{11} to 10^{12} ions/cm^2 and is used in the bombardment of monolayers lying closest to the surface. In the dynamic mode, a dose of ions stronger than 10^{13} ions/cm^2 is used, and results in a higher number of excited atoms. High current density of primary ions causes continuous changes in material under study (the implantation of ions, deterioration of the crystal lattice, changes in the chemical composition), so this mode is not suitable for the characterization of monolayers of a solid surface, only to perform depth profiles (e.g., alloys, glasses).

TOF-SIMS Modes of Operation

The TOF-SIMS spectrometer offers studies in three main modes:

1. *Surface spectroscopy* (*static SIMS*): investigation of the chemical composition of a surface through its mass spectrum
2. *Surface imaging*: investigation of the surface distribution of elements and compounds by obtaining mass resolved secondary ion images (chemical maps)
3. *Depth profiling* (*dynamic SIMS*): investigation of the chemical composition of the inside of a sample by obtaining the distribution of elements and small clusters as a function of depth

Static SIMS   A basic problem in research of a large area of a surface layer is determination of the chemical composition of the sample surface with the best sensitivity. To reach this aim, the beam of a very low dose of primary ions irradiates an area 0.01 mm to several mm^2 in size. Secondary ions emitted from the surface analyzed are presented in the form of a mass spectrum, which makes it possible to define the entire composition of the sample surface. A TOF-SIMS experiment provides a mass spectrum from the uppermost 10 to 20 Å of a sample, which is a composite of mass spectra from each of the species present on the surface. It is assumed that 1 mm^2 of the monolayer area contains about 5×10^{12} surface particles. The lowest concentration it is possible to detect is about 1 ppm.

This method is widely used to characterize all types of surfaces important from a technological point of view, subjected to, among other things, friction and corrosion, and in research on the surface adhesion proprieties, surface modification, and identification of defects and detection of trace levels of impurities.

Surface Imaging   Surface imaging with chemical analysis of microareas can be achieved by rastering of a finely focused ion beam over a very small surface (reduction in the diameter of the primary ion beam). As a result, maps of secondary ion images are generated. Scanning by so well focused an ion beam over a sample surface (Figure 3) makes it possible to study the chemical composition of microareas, and consequently, enables determination of atom or particle distribution on the surface (investigation of surface heterogeneity). It should be noted that in a TOF-SIMS imaging experiment, a mass spectrum is collected from each pixel in the image. Spatial resolution of the

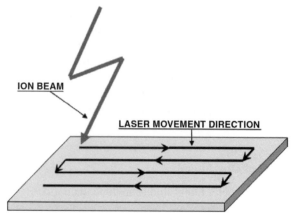

FIGURE 3 Direction of plate scanning.

method refers to the smallest distance between two points on the sample surface that can be distinguished and is described by the diameter of the primary beam, which can vary from 50 nm to several millimeters.

TOF-SIMS Imaging Application

The development and introduction of a new generation of ion sources, such as poly-atomic clusters and SF$_5$, C$_{60}$, Au, and Bi clusters, greatly enhance ion and molecular mass yields, which is especially important for molecular SIMS imaging as well as for molecular depth profiling. TOF-SIMS imaging has become a very attractive method for the localization of unknown molecules at sample surfaces, primarily in the analysis of organic, forensic, medical, and biological samples (e.g., tissue sections, cells, biomaterials, pharmaceuticals).

Figure 4 presents ion images obtained from a rat brain section under the irradiation of Bi$_3^+$ primary ions corresponding to the positive fragment ions of (A) choline (*m/z* 86), (B) a phosphocholine head group (*m/z* 184), (C) cholesterol [M + H − H$_2$O]$^+$ (*m/z* 369), (D) [M − H]$^+$ (*m/z* 385), (E) vitamin E (*m/z* 430), and (F) phospholipids (mean *m/z* 769). Figures 5 to 7 show examples of an application of TOF-SIMS imaging for the forensic examination of fingerprints. Preliminary studies were focused on chemical investigations of fingerprints and detection traces of substances that do not exist in natural excretion but derive directly from crime: for example, Ni, gunpowder residues, and arsenic. This domain of study is very important and can result in a new type of dactyloscopy. TOF-SIMS imaging may also be expected to be useful in solving other problems of forensic science (e.g., fingerprint deposit transfer, fingerprint aging process, gunshot residues, or fiber examination). Figure 5 presents an image of a whole fingerprint obtained from a stainless-steel sheet surface. The large image size (26,000 × 26,000 μm^2) results in characteristic fingerprint lines not being sufficiently visible. Cr, Mn, and Fe ions originate from a stainless-steel surface, whereas Na, Ca, K, Si, Si-oil (mass 73), or Si-oil (mass 147) come from a finger as a natural excretion.

Field of view: 18000.0 × 18000.0 μm²

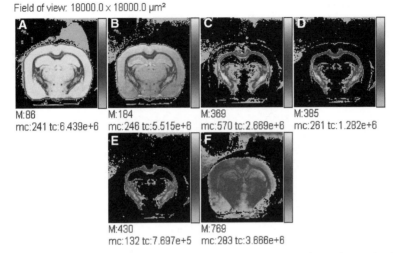

M:86
mc:241 tc:6.439e+6   M:184
mc:246 tc:5.515e+6   M:369
mc:570 tc:2.669e+6   M:385
mc:261 tc:1.282e+6

M:430
mc:132 tc:7.697e+5   M:769
mc:283 tc:3.666e+6

FIGURE 4   Positive secondary ion images obtained from a rat brain section under the irradiation of Bi_3^+ primary ions. 256×256 pixels, pixel size 70×70 μm². [Reprinted from Touboul et al. (2005) with permission from Elsevier.] (*See insert for color representation of figure.*)

Field of view: 26000.0 × 26000.0 μm²

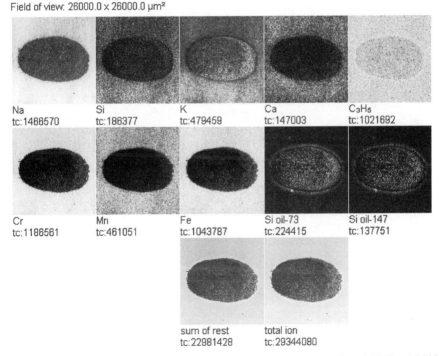

Na
tc:1466570   Si
tc:186377   K
tc:479459   Ca
tc:147003   C₃H₅
tc:1021692

Cr
tc:1186561   Mn
tc:461051   Fe
tc:1043787   Si oil-73
tc:224415   Si oil-147
tc:137751

sum of rest
tc:22981428   total ion
tc:29344080

FIGURE 5   Image of a fingerprint taken from a stainless-steel sheet surface ($26{,}000 \times 26{,}000$ μm²). (*See insert for color representation of figure.*)

Field of view: 10000.0 × 10000.0 μm²

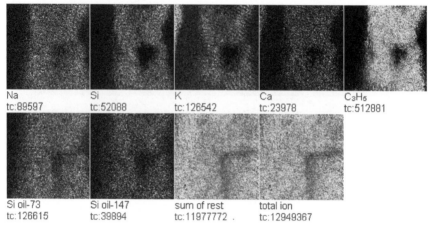

FIGURE 6   Image of a fingerprint taken from a paper surface (10,000 × 10,000 μm²). (*See insert for color representation of figure.*)

Field of view: 10000.0 × 10000.0 μm²

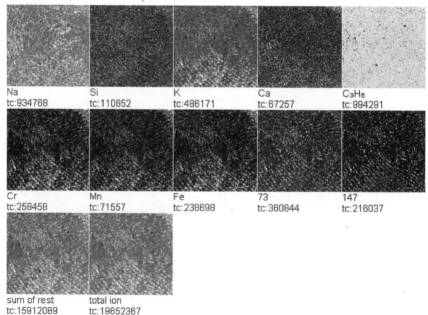

FIGURE 7   Image of a fingerprint taken from a stainless-steel sheet surface (10,000 × 10,000 μm²). [Reprinted from Szynkowska et al. (2007) with permission.] (*See insert for color representation of figure.*)

Some characteristic areas of fingers, called loops, spirals, whirls, tents, and so on, can be observed in the images of fingerprints taken from paper and stainless-steel surfaces from a field of view of 10,000 × 10,000 μm² (Figures 6 and 7).

MALDI-TOF

Historically, MALDI-TOF was the first MS technique (after TOF-SIMS, which was dedicated to scanning purposes) adopted to the imaging requirements. At that time there was an urgent need for software development, both for instrument control and for data collection and evaluation. As in its "classic" form, MALDI-TOF operates at high vacuum, many technical problems soon emerged. First, biological tissue contains plenty of water. Pumping it out of a tissue leads to shrinking of the tissue and generates irreproducible data. Fast scanning also demands novel, faster laser, and this has already been developed by manufacturers, which now offer several types of laser technologies (UV, infrared). Another, even more serious problem was application of the matrix. Today, several techniques for matrix application are available, improving the general usefulness at least partially, but the method, although used extensively by various laboratories, is still far from routine. Figure 8 is an example of MALDI imaging.

To visualize the distribution of many molecules requires good resolution of the analysis, but this depends primarily on the dimensions of the laser beam. MALDI-TOF can achieve spatial resolution below 50 μm (often around 10 μm). Such fine

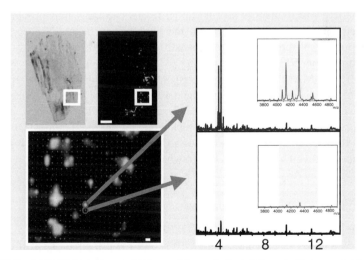

FIGURE 8 MALDI imaging at high spatial resolution (50 μm). Beta-amyloid peptide plaques in Alzheimer mouse brain section. (Courtesy of M. Riemenschneider, Klinikum Rechts der Isar, TU-München, Munich, and Bruker Daltonics, Bremen, Germany. (*See insert for color representation of figure.*)

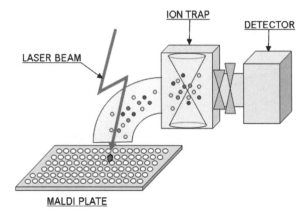

FIGURE 9   AP- MALDI. (*See insert for color representation of figure.*)

sampling can generate a huge amount of data, sometimes exceeding the capacity of the hard disk. Therefore, this parameter should be considered carefully when selecting optimal strategy.

Atmospheric Pressure–MALDI

Atmospheric pressure (AP)–MALDI works in a manner similar to MALDI (Figure 9). The substantial difference is inclusion of an infrared laser, which does not require a matrix for sample ionization. Another interesting feature is that at atmospheric pressure there is no substantial tissue shrinking, thus providing better reproducibility. As no matrix is used, the method can be used successfully for detection of small molecules, which usually are obscured by the highly abundant matrix ions below 500 to 700 m/z.

Usually, AP-MALDI source is mounted on an ion-trap instrument by replacement of the electrospray device. Therefore, the mass range of the measurements is limited to the working range of the ion-trap analyzer (ca. 2000 to 6000 m/z), bearing in mind all problems associated with this technique not observed in the TOF analyzer. On the other hand, an ion trap provides an unprecedented possibility for multiple fragmentations, which overcomes the drawbacks noted above.

Desorption by Electrospray Ionization

Desorption by electrospray ionization (DESI), the youngest technique, may also be utilized as a molecular scanner. Figure 10 shows the principle of DESI operation. Samples deposited on a flat target (e.g., for MALDI) are ionized and desorbed by electrospray. The method is highly versatile, robust, and does not require a specific matrix or solvent to be used. It is also more tolerant to salt, which is a common obstacle when using other ionization techniques. The major drawback to be solved is

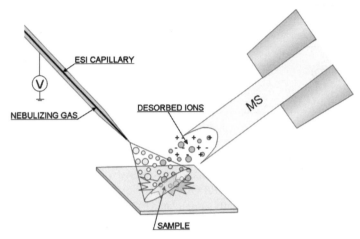

FIGURE 10   DESI operational principle. (*See insert for color representation of figure.*)

lower resolution than for a laser beam. Some recent attempts aim at better focusing of the spray by using nanoelectrospray.

Several technical aspects need to be considered during DESI operation:

- Collection angle between inlet capillary and target plate
- Incident angle between target plate and spray
- Distance between target plate and inlet capillary
- Spray tip diameter

A variant of DESI has been developed involving desorption atmospheric pressure chemical ionization (DAPCI), where gaseous ions of volatile compounds such as toluene can be used for ionization. The major difference between DESI and DAPCI is installation of corona discharge in the latter (i.e., construction resembling an APCI system).

Other, less commonly used techniques are:

- Direct analysis in real time (DART)
- Electrospray-assisted laser desorption/ionization (ELDI)
- Atmospheric solids analysis probe (ASAP)
- Jet desorption ionization (JeDI)
- Plasma-assisted desorption ionization (PADI)

Details of their operation and possible applications may be found in the review by Cooks et al. (2006). Some of the most important features of the methods described here are collected in Table 1.

TABLE 1   Comparison of Ionization Techniques Used for Imaging MS

Method	Mass Range	Resolution (μm)
TOF-SIMS	50–10,000	10–50
MALDI-TOF	500–700,000	10–50
AP-MALDI	50–6000	10–50
DESI	50–6000	300

Acknowledgments

This work was supported by International Centre for Genetic Engineering and Biotechnology (ICGEB) grant CRP/POL05-02. The assistance of M. Riemenschneider and Bruker Daltonics is greatly appreciated.

RECOMMENDED READING

Belu M., Graham D.J., Castner D.G. Time-of-flight secondary ion mass spectrometry: techniques and applications for the characterization of biomaterial surfaces. *Biomaterials* 21 (2003) 3635–3653.

Benninghoven A., Rudenauer F., Werner H.W. *Secondary Ion Mass Spectrometry*. Wiley, New York, 1987.

Cooks R.G., Ouyang Z., Takats Z., Wiseman J.M. Ambient mass spectrometry. *Science* 311 (2006) 1566–1570.

Grams J., Szynkowska M.I., Norris C.P. eds. Applications of time-of-flight secondary ion mass spectrometry (TOF-SIMS) to investigations of metal/support catalysts. In *Focus on Surface Science Research*. Nova Science Publishers, Hauppauge, NY, 2005, pp. 233–263.

Pacholski M.L., Winograd N. Imaging with mass spectrometry. *Chem. Rev.* 99(1999) 2977–3005.

Paryjczak T., Szynkowska M.I. The use of time-of-flight secondary ion mass spectrometry (TOF-SIMS) for solid surface studies. *Przem. Chem.* 82(3) (2003) 199–206.

Szynkowska M.I. *Imaging of Small Molecules: Mass Spectrometry for Students*. Wiley, Hoboken, NJ, in press.

Szynkowska M.I., Czerski K., Grams J., Paryjczak T., Parczewski A. Preliminary studies using scanning mass spectrometry (TOF-SIMS) in the visualisation and analysis of fingerprints. *Imag. Sci. J.* 55(2007) 180–187.

Takats Z., Wiseman J.M., Cooks R.G. Ambient mass spectrometry using desorption electrospray ionization (DESI): instrumentation, mechanisms and applications in forensics, chemistry and biology. *J. Mass Spectrom.* 40(2005) 1261–1275.

Touboul D., Kollmer F., Niehuis E., Brunelle A., Laprevote O. Improvement of biological time-of-flight secondary ion mass spectrometry imaging with a bismuth cluster ion source. *J. Am. Soc. Mass Spectrom.* 16(2005) 1608–1618.

Vickerman J.C., Briggs D. *TOF-SIMS Surface Analysis by Mass Spectrometry*. Surface Spectra/IMP, Chichester, UK, 2001.

7.4

FRAGMENTATION AND SEQUENCING IN PROTEOMICS

PIOTR SUDER

USE OF TANDEM MASS SPECTROMETRY IN PROTEOMICS

The proteomic approach differs significantly from the "classical" identification of a single protein or peptide as it concerns simultaneous, large-scale analysis in a relatively short time. Such a procedure must be automated, at least partially. This, in turn, indicates that the sequencing step needs to be performed without an operator's intervention, based on the software and on the protein databases. The routine used in the proteomics approach also differs in terms of the algorithms applied. Either peptide maps serve for identification or peptide tags are loaded directly to MASCOT or ProteinProspector, for example, and analyzed automatically. Sequencing of individual peptides with the help of tandem mass spectrometry is based on thorough analysis of the mass spectrum, without additional software. Therefore, it is advisable to acquire knowledge of the type of fragment formed during MS/MS analysis and on the basics of a fragmentation mechanism, as there is often a need to verify manually the data obtained, or to discover a novel sequence which is not yet present in the database. Details on peptide fingerprinting are described in Chapter 7.5, and below we provide a brief introduction to de novo sequencing, which is an exciting challenge to everyone dealing with the discovery of novel molecules. Due to the limited space and the broadness of the topic, we strongly recommend that readers consult additional literature, such as the Matrix Science Web page: http://www.matrixscience.com/help_index.html.

The ionization modes commonly used in proteomics, electrospray ionization (ESI), and matrix-assisted laser desorption/ionization (MALDI) belong to the category of "soft" ionization methods. Therefore, almost no spontaneous fragmentation is

Proteomics: Introduction to Methods and Applications, Edited by Agnieszka Kraj and Jerzy Silberring
Copyright © 2008 John Wiley & Sons, Inc.

FIGURE 1 MS/MS procedure.

observed. Forced fragmentation, also called *tandem mass spectrometry*, is one of the most informative techniques used in mass spectrometry for substance identification. The entire process has three steps:

1. Isolation of the ion of interest
2. Ion dissociation
3. Separation and analysis of the fragments

Based on the data obtained, it is possible to reveal at least partial amino acid sequences of peptides or to verify the structure of smaller organic compounds. The latter can also serve, in conjunction with high-resolution mass spectrometry, as a tool for the determination of elemental composition. The entire procedure is shown in Figure 1.

The first step, isolation, is necessary to select and isolate an ion of interest from a mixture of other ions present in the mass spectrum. In general, it is possible to analyze one ion at a time. The selection/isolation process occurs in the collision chamber (e.g., quadrupole, ion-trap, ion cyclotron). The ion selected is referred to as a *parent ion.*

The fragmentation process involves an inert collision gas (He, Ar), which enters the collision chamber and collides with sample molecules. It is now clear that the pressure (quantity) of the collision gas molecules will determine the efficiency of fragmentation process. An insufficient amount of gas will lead to ineffective fragmentation. In contrast, too high gas pressure will dampen all ions.

Daughter ions, generated during collisions, are separated and analyzed in a third quadrupole, an ion trap, or an ion cyclotron, for example. The most important conclusion is that all ions (i.e., fragments or daughter ions) formed during the previous phase belong to the parent ion selected. This, in turn, leads to easier sequencing or structure verification.

The isolation, fragmentation, and separation steps may be repeated n times to receive MS^n spectra, where n indicates the number of cycles. In modern instruments it is possible to perform up to 11 or 12 repeats, but in practice, two or three cycles are generally used. Usually, when a complex mixture is analyzed with the aid of an LC-MS technique, there is not enough time to perform more than MS^3, which is usually sufficient for partial sequence identification. Peaks emerging from the capillary column are very narrow, lasting about 15 to 30 seconds, and during this short time, the full scan, MS^2, and MS^3 should be completed, including switching between the modes, ion selection, and return to full scan.

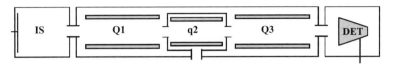

FIGURE 2   Triple quadrupole analyzer: IS, ion source; Q1, quadrupole 1; q2, collision chamber; Q3, quadrupole 2; DET, detector.

Fragmentation Strategies

Two fragmentation strategies are commonly used, depending on the MS construction: Tandem in space and tandem in time.

Tandem in Space   The simplest way to perform fragmentation is to use two analyzers separated by a collision chamber. An example is an instrument equipped with a triple-quadrupole analyzer (Figure 2). In an MS/MS experiment, the ions formed in the ion source travel through the first analyzer (Q1). It is possible to adjust this device as a mass filter, thus allowing ions with a unique m/z ratio to pass the quadrupole. Only this ion will be fragmented in the next quadrupole, acting as a collision chamber. The parent ion selected is introduced into the collision chamber (q2). Here, an inert gas (argon or helium) collides with the parent ion. As an effect of these collisions, the parent ion dissociates into fragments. The fragments obtained (daughter ions) are guided by the ion optics to the second analyzer (Q3), which separates them according to their m/z ratios, similar to a conventional MS experiment.

Tandem in Time   Another type of instrumentation utilizes a quadrupole ion trap (IT) or ion cyclotron resonance (ICR) analyzer, which collects ions, isolates them, and carries out fragmentation and analysis of daughter ions in the same chamber (space). During the first phase, the ion trap is filled with ions generated in the ion source. This analyzer can collect the ions and keep them at stable orbits. It is possible to perform up to 11 cycles of fragmentation in such devices, although in fact, such capabilities are very seldom used, and for proteomics purposes, fragmentation above MS^3 is meaningless. To resolve structures of complex molecules, MS^6 or MS^7 is often used. Further details describing the construction and principles of operation of the instrument described above may be found in the recommended reading.

Earlier we discussed fragmentation linked to ES ionization, which is still the approach in most common use. Older MALDI-TOF constructions had an option called post-source decay (PSD), which is not fully efficient during fragmentation and requires a reflectron. A more advanced technical solution involves MALDI ionization and a TOF analyzer equipped with an IT (MALDI-IT-TOF). The principle of operation is shown in Figure 3. The ion trap, installed between the ion source and an analyzer, serves as an ion separation device in the MS^n mode, when n is higher than 2. The main disadvantages of this setup are the limited mass range and decreased resolution, usually not higher than 3000 to 4000. Alternatively, the IT can be disconnected, and the instrument works as a normal MALDI-TOF device. Such

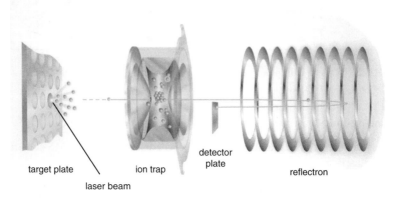

FIGURE 3   MALDI-IT-TOF mass spectrometer. (From Shimadzu Corp. application note MO185, 2nd ed.)

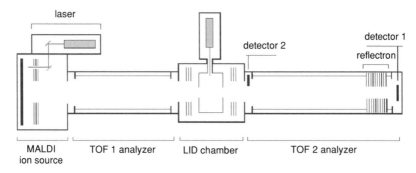

FIGURE 4   TOF/TOF instrument equipped with a laser-induced dissociation chamber.

a fusion of two analyzers in a single instrument also permits very high sensitivity toward ions generated in the instrument.

An alternative use of a TOF/TOF analyzer in which a complete amino acid sequence may be obtained is shown in Figure 4.

The capability of MALDI-TOF instruments to perform efficient, high-throughput fragmentation significantly extends the number of methods utilized in proteomics.

FRAGMENTATION PROCESS

A molecule requires additional external energy to break down covalent bonds. This additional energy can be provided using one of several techniques, the most popular being:

1. Low-energy collision-induced dissociation (CID)
2. High-energy CID

3. Dissociation of metastable ions
4. Electron capture dissociation (ECD)

Low-Energy CID

In this technique the energy necessary for dissociation comes from abundant collisions of parent ions with inert gas molecules. The collision energy is not higher than 100 to 200 eV. This method is used generally in all analyzers linked to ESI and MALDI sources (i.e., triple quadrupole, IT, ICR). Such collisions produce a series of fragments, mostly of the type of b and y''. This feature is important during de novo sequencing, as we may predict ion types depending on the instrument used.

High-Energy CID

Similar to low-energy CID, fragmentation utilizes collisions with inert gas molecules. Because of the much higher energy of the ions, the collision energy is much higher and achieves 5 to 10 keV in many cases (depending on the acceleration voltage used in a particular instrument). This type of fragmentation is achievable in sector instruments. The MALDI-IT-TOF instrument mentioned above is equipped with an ion source operating at high energy; therefore, to fully utilize IT capabilities, the ions leaving the source must be cooled before they enter the ion trap. In some instruments, the high energy of ions is increased additionally by irradiation by a laser beam [called laser-induced dissociation (LID)] commonly used in TOF/TOF analyzers. The latter type of fragmentation generates a, b, y, and i ions.

Detection of Metastable Ions

A metastable ion (in mass spectrometry) is an ion formed with sufficient excitation to dissociate spontaneously during its flight from the ion source to the detector (*IUPAC Compendium of Chemical Terminology*). In some cases, ions leaving an ion source possess energy sufficient for spontaneous fragmentation without the presence of collision gas. Such ions, the metastable ions, dissociate spontaneously while traveling through an analyzer resulting in a fragmentation spectrum. During the dissociation process, the fragments have identical velocity but different kinetic energies. Such fragmentation is commonly detected in sector instruments and also in MALDI-TOF mass spectrometers. The latter utilizes a post-source decay procedure for the detection of such ions.

Electron Capture Dissociation

Electron-capture dissociation fragmentation utilizes a different mechanism from the others that we have discussed. The ions do not collide with an inert gas; instead, they interact with a low-energy electron beam. The ions recombine with electrons and their charge is reduced to $n - 1$. Formation of such an ion containing an odd electron leads to its decomposition (fragmentation). Therefore, this technique is only suitable for multiply charged ions (e.g., produced by ESI). This type of fragmentation is also suitable for large molecules such as proteins (top-down proteomics) or nucleic acids. Not only are the weakest bonds destroyed, but the molecules can be fragmented more efficiently, so this technique provides much more information than does CID. A disadvantage is the long time necessary for fragmentation; thus, only analyzers capable of retaining the ions longer are suitable. Moreover, low-energy electrons require high vacuum, and an ion trap working under moderate pressure is ineffective. This type of ionization is often used in ion cyclotron resonance (ICR) and IT analyzers. As an alternative to ECD, a similar method has been introduced, which electron transfer dissociation, utilizes anions (e.g., anthracene) instead of electrons. This method produces mainly c and z ions and is particularly useful for the identification of posttranslational modifications (see also Chapter 8).

Infrared Multiphoton Dissociation

Infrared multiphoton dissociation (IRMPD) involves use of an infrared laser. Ions formed during this process absorb multiple infrared photons and undergo fragmentation. Although IRMPD involves low-energy photons, it deposits energy into the precursor and product ions, whereas CID acts only on precursors. IRMPD creates mostly b and y ions, but it is more pronounced than CID fragmentation, because fragment ions generated by IRMPD are further irradiated by the laser and also fragmented. IRMPD is more efficient for larger ions, such as peptides, proteins, and carbohydrates. This technique allows for detailed analysis of phosphopeptides, disulfide bridges, acetylation, and other posttranslational modifications. Because of the technique specificity and the time required for laser irradiation, ion trapping instruments (e.g., ICR or ion trap) are most suitable.

NOMENCLATURE OF FRAGMENTATION SPECTRA

In the early 1980s, Roepstorff and Fohlmann (1984) proposed an unequivocal system describing fragment ions of peptides. It was modified slightly by Johnson (1987) and used widely until at present. In the system, ions are marked using one letter and one number. The simplest way to understand the system is to analyze Figure 5.

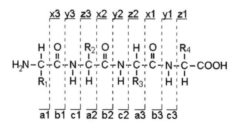

FIGURE 5   Roepstorff's and Fohlman's fragment ion nomenclature.

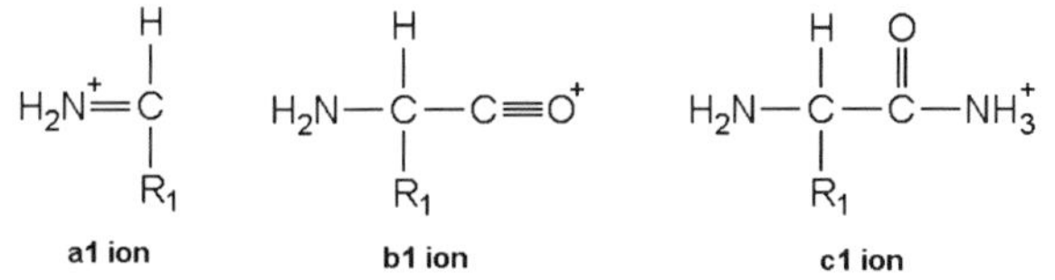

$$H_2N\overset{+}{=}\underset{R_1}{\overset{H}{\underset{|}{C}}} \qquad H_2N-\underset{R_1}{\overset{H}{\underset{|}{C}}}-C\equiv O^+ \qquad H_2N-\underset{R_1}{\overset{H}{\underset{|}{C}}}-\overset{O}{\overset{\parallel}{C}}-NH_3^+$$

a1 ion   **b1 ion**   **c1 ion**

FIGURE 6 Examples of ions formed named according to Roepstorff's and Fohlman's nomenclature.

Ions from the N-terminus are named using letters from the begenning of the alphabet (i.e., a, b, c). The digit indicates the number of residues present in the ion. C-terminal fragments are labeled using letters from the end of the alphabet (i.e., z, y, x). Examples are shown in Figure 6.

This nomenclature becomes complicated, when an ion loses an additional —OH group and a single proton (called H_2O loss) or —NH_2 group (amino group loss). This happens particularly in spectrometers equipped with a quadrupole or quadrupole ion-trap analyzer. Daughter ion formed after water loss are labeled with a degree sign (e.g., $a°$, $y°$), and amino group loss is marked by an asterisk (e.g., a^*, y^*). Quite often, fragmentation occurs in ions that have a charge higher than +1 or −1. In such cases, additional charge should be included in the label. A few examples of daughter ion nomenclature are given in Figure 7.

Immonium ions are internal fragments derived from a single amino acid residue by a combination of a and y-type cleavage, often formed during high-energy collisions. The typical structure of such ions is shown in Figure 8. They are labeled using one-letter codes derived from their corresponding amino acids (Table 1) and are important diagnostic ions that confirm the presence of a particular residue. As always, it should be emphasized that the absence of such an ion does not indicate that the amino acid is not present.

It is important to remember that almost every bond in a peptide structure may be broken upon fragmentation. One example is the formation of satellite ions, derived from the ionized side-chain groups of amino acid residues. This process occurs during high-energy CID. A typical example is the possibility to distinguish between

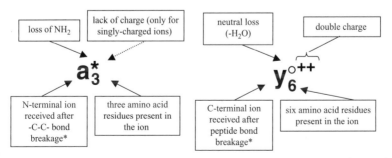

FIGURE 7 Typical examples of daughter ion nomenclature (for ions marked a^*_3 and $y°^{++}_6$). (*See Figure 5.)

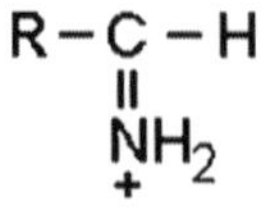

$$R-\underset{\underset{+}{\overset{\|}{NH_2}}}{C}-H$$

FIGURE 8   Typical structure of an immonium ion.

TABLE 1   Immonium and Related Ion Masses[a]

Amino Acid	Immonium Ions (m/z)	Derivatives (m/z)
Alanine (Ala, A)	44	
Arginine (Arg, R)	*129*	59,70,73,87,100,112
Asparagine (Asn, N)	*87*	*70*
Aspartic acid (Aps, D)	*88*	*70*
Cysteine (Cys, C)	*76*	
Glutamic acid (Glu, E)	*102*	
Glutamine (Gln, Q)	101	56,84,129
Glycine (Gly, G)	30	
Histidine (His, H)	**110**	*82,121,123,138,*166
Isoleucine (Ile, I)	**86**	44,72
Leucine (Leu, L)	**86**	44,72
Lysine (Lys, K)	*101*	70,84,112,129
Methionine (Met, M)	*104*	61
Phenylalanine (Phe, F)	**120**	*91*
Proline (Pro, P)	**70**	
Serine (Ser, S)	60	
Threonine (Thr, T)	74	
Tryptophan (Trp, W)	**159**	77,117,**130**,132,**170,171**
Tyrosine (Tyr, Y)	**136**	*91,107*
Valine (Val, V)	72	41,55,69

Source: Falick (1993) and Papayannopoulos (1995).
[a]Bold letters indicate abundant signals; italic letters indicate weak intensity.

the isobaric amino acids isoleucine and leucine. The ions are also useful during the search for mutations in peptide or protein sequences. Such low-mass ions may be missed when an ion trap is used as an analyzer. This device has a low-mass cutoff during MS/MS experiments, and thus we can lose additional important information on diagnostic ions.

SUMMARY

Fragmentation is used in proteomics mainly to determine sequences of amino acid residues in peptides, which leads to identification of the peptide or protein of interest. This technique is much faster, more sensitive, and less expensive that Edman sequencing, so is slowly replacing that classical identification method. Fragmentation capabilities of mass spectrometers are used primarily in top-down and bottom-up proteomics. In the top-down methodology, the entire protein or its large fragments are subjected to fragmentation in a mass spectrometer. In the bottom-up approach, the protein is cleaved by a proteolytic enzyme of known specificity (usually, trypsin) and then these tryptic peptides are fragmented (see also Chapters 7.5 and 7.6).

Acknowledgments

This work was supported by International Centre for Genetic Engineering and Biotechnology (ICGEB) grant CRP/POL05-02.

SELF-STUDY QUESTIONS

1. During peptide sequencing, after analyzing the mass spectra obtained, we can reconstruct the sequence of amino acid residues in the peptide chain. How is this possible?
2. Why do we usually not fragment large proteins using an electrospray technique? Is it possible to carry out an MS/MS experiment on large molecules?
3. Leucine and isoleucine amino acid residues seem to be indistinguishable using an MS/MS technique because of having the same molecular mass. Is it possible to plan an MS/MS experiment that is able to distinguish these residues?
4. Is it possible to do MS^3 using a mass spectrometer equipped with a triple-quadrupole analyzer? If so, what conditions with regard to the sample and the ion source must be fulfilled to obtain an MS^3 spectrum?

RECOMMENDED READING

Bogdanov B., Smith R.D. Proteomics by FTICR mass spectrometry: top down and bottom up. *Mass Spectrom. Rev.* 24(2005) 168–200.

Chapman J.R. *Practical Organic Mass Spectrometry*, 2nd ed. Wiley, New York, 1995.

Johnstone R.A.W., Rose M.E. *Mass Spectrometry for Chemists and Biochemists*, 2nd ed. Cambridge University Press, New York, 1996.

Udeshi N.D., Shabanowitz J., Hunt D.F., Rose K.L. Analysis of proteins and peptides on a chromatographic timescale by electron transfer dissociation MS. *FEBS J.* 274(2007) 6269–6276.

Wolf-Yadlin A., Hautaniemi S., Lauffenburger D.A., White F.M. Multiple reaction monitoring for robust quantitative proteomic analysis of cellular signalling networks. *Proc. Natl. Acad. Sci. U.S.A.* 104(2007) 5860–5865.

Wysocki V.H., Resing K.A., Zhang Q., Cheng G. Mass spectrometry of peptides and proteins. *Methods* 35(2005) 211–222.

7.5

PROTEIN IDENTIFICATION BY MASS SPECTROMETRY

Anna Drabik, Marek Noga, Marcus Macht, and Jerzy Silberring

PEPTIDE MASS FINGERPRINTING

Peptide mass fingerprinting (PMF; also known as *peptide mass mapping*) is a technique that allows for protein identification by mass spectrometry (MALDI or ESI) as a result of matching their peptide fragments to the theoretical peptide masses generated from databases through use of a mass search program. As this technique relies on the investigation of a protein digest rather than direct analysis of the protein itself, it is a bottom-up protein identification strategy, as presented in Figure 1.

The first step in peptide mass analysis is always a more or less selective cleavage by a proteolytic enzyme (e.g., trypsin, chymotrypsin, pepsin, carboxypeptidases, thrombin, elastase) or chemical agent (e.g., CNBr, HCl, TFA). However, to achieve reliable and comparable digestion, the sample has to be treated prior to proteolysis. Reduction with dithiothreitol (DTT) or mercaptoethanol, for example, causes breakage of disulfide bridges between cysteine residues. Subsequently, alkylation of thiol groups by iodoacetamide is required to prevent cysteine residues resulting from the spontaneous re-formation of disulfide bridges (see Chapter 3.1).

After digestion, the peptides generated, which constitute a peptide map, are analyzed by mass spectrometry. The basic concept of PMF is that each protein has a unique set of peptides and consequently, unique peptide masses. The series of peptide fragments is specific for only one protein and is referred to as a peptide fingerprint. Identification is done by comparing experimental peptide masses with theoretical

Proteomics: Introduction to Methods and Applications, Edited by Agnieszka Kraj and Jerzy Silberring
Copyright © 2008 John Wiley & Sons, Inc.

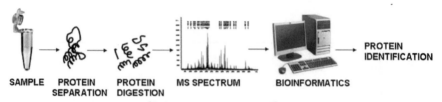

SAMPLE   PROTEIN   PROTEIN   MS SPECTRUM   BIOINFORMATICS
          SEPARATION DIGESTION

FIGURE 1   PMF procedure.

digestion of all protein sequences from a sequence database (e.g., SwissProt, MSDB, NCBI). The entire process is performed using mass search software (e.g., MASCOT, ProFound, ProteinProspector, Proteomics, the sequence analysis tool Expasy).

The reliability of identification by this method increases with the number of peptide masses identified and the measurement accuracy. PMF fails to identify protein mixtures and is very sensitive to contamination with other proteins, such as keratins (a group of proteins present in dust, skin, and clothes). Their presence obscures the results because the PMF algorithm cannot distinguish which peptide is derived from which protein; therefore, during the fitting process some fragments are considered as contaminants and result in a reduced score number. Therefore, PMF is often preceded by preseparation of proteins to ensure the purity of the peptide map. Also, special precautionary steps are taken to avoid or at least to minimize contamination by keratins. PMF identification is much more reliable when combined with additional information gained through electrophoretic (molecular mass) or chromatographic (retention time) separation.

The most commonly used technique suitable for PMF is MALDI-MS (Chapter 7.2), as it is fast, robust, accurate, sensitive, easy to perform, and can be automated. Alternatively, ESI (Chapter 7.1), in combination with LC, can be used to achieve peptide mass information.

MASCOT, MS-Fit, and ProFound are the most commonly used Internet-accessible engines for PMF. Generally, the same results should be obtained from different software using identical parameters. Even though PMF is rather simple for an operator, various factors may influence the results of the analysis. Specific issues and features will be discussed in detail based on the MASCOT search engine.

After MALDI-MS analysis, a list of peptide masses is generated and transferred into the PMF search program (Figure 2). The NCBI and MSDB databases are larger; SwissProt has fewer entries. However, SwissProt provides all information for each protein manually verified and annotated, resulting in much more reliable and detailed data.

The list of peptide masses may be compared against the entire database, and in such a case the taxonomy should be set to "all entries." If the origin of the sample is known, the organisms being sought can be specified. This can reduce the search time significantly by decreasing the number of potential results, and may result in higher specificity.

In addition, the name of the enzyme used for proteolytic digestion needs to be specified (Figure 3), as well as the number of missed cleavage sites to consider.

MASCOT Peptide Mass Fingerprint

Your name		Email	
Search title			
Database	MSDB		
Taxonomy	All entries		
Enzyme	Trypsin	Allow up to	1 ▼ missed cleavages

Fixed modifications	Acetyl (K) Acetyl (N-term) Acetyl (Protein N-term) Amidated (C-term) Amidated (Protein C-term)	Variable modifications	Acetyl (K) Acetyl (N-term) Acetyl (Protein N-term) Amidated (C-term) Amidated (Protein C-term)

Protein mass _____ kDa   Peptide tol. ± 1.2   Da ▼

Mass values ● MH⁺ ○ Mᵣ ○ M-H⁻   Monoisotopic ● Average ○

Data file _____   [Search...]

Query
NB Contents
of this field
are ignored if
a data file
is specified.

Decoy ☐   Report top AUTO ▼ hits

[Start Search ...]   [Reset Form]

FIGURE 2   MASCOT peptide mass fingerprinting interface. (From http://matrixscience. com/cgi/search_form.pl?FORMVER=2&SEARCH=PMF.)

The enzyme used most frequently is trypsin, because of its high purity, selectivity, activity, stability, and ability to perform enzymatic digestion within a gel matrix. The fact that peptides released by trypsin digestion possess at least one basic amino acid is important, as it improves the ionization process by localization of the charge. Setting the number of missed cleavages to 0 or 1 reduces the number of peptides matched to the experimental data. Peptides obtained from autolysis of the proteolytic enzyme

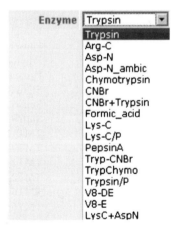

Enzyme [Trypsin ▼]
Trypsin
Arg-C
Asp-N
Asp-N_ambic
Chymotrypsin
CNBr
CNBr+Trypsin
Formic_acid
Lys-C
Lys-C/P
PepsinA
Tryp-CNBr
TrypChymo
Trypsin/P
V8-DE
V8-E
LysC+AspN

FIGURE 3   Choice of digestion method. (From http://matrixscience.com/cgi/search_form. pl?FORMVER=2&SEARCH=PMF.)

are also present on the MS spectra. For matching, only a few good-quality peaks (high intensity, presence of isotopic distribution) should be included, to minimize the noise (a higher score will be generated by identifying 4 of 5 peptides submitted for database query rather than 4 of 15).

When iodoacetamide is used to modify cysteine groups, carbamidomethyl has to be set up as a fixed modification (Table 1). In the absence of iodoacetamide, propionamide (C) modification should be chosen. The most common variable modifications are acetyl (protein), pyro-Glu (N-term Q), and oxidation (M). The rationale is that oxidation is spontaneous, and both types of methionine groups should be considered: oxidized as well as reduced forms. For initial searches, however, it is recommended that only methionine oxidation be considered as a

TABLE 1   Fixed and Variable Amino Acid Modifications

Acetyl (K)	ICPL:2H(4) (protein N-term)
Acetyl (N-term)	iTRAQ4plex (K)
Acetyl protein (N-term)	iTRAQ4plex (N-term)
Amidated (C-term)	iTRAQ4plex (Y)
Amidated protein (C-term)	iTRAQ8plex (K)
Ammonia-loss (N-term C)	iTRAQ8plex (N-term)
Biotyn (K)	iTRAQ8plex (Y)
Carbamidomethyl (C)	Label:18O(1) (C-term)
Carbamyl (K)	Label:18O(2) (C-term)
Carbamyl (N-term)	Met→Hse (C-term M)
Carboxymethyl (C)	Met→Hsl (C-term M)
Cation:Na (C-term)	Methyl (C-term)
Cation:Na (DE)	Methyl (DE)
Deamidated (NQ)	Methylthio (C)
Dehydrated (N-term C)	NIPCAM (C)
Dehydro (C)	Oxidation (HW)
Dioxidation (M)	Oxidation (M)
Ethanolyl (C)	Phospho (ST)
ExactagAmine (K)	Phospho (Y)
ExactagThiol (C)	Propionamide (C)
Formyl (N-term)	Pyrocarbamidomethyl (N-term C)
Formyl (protein N-term)	Sulfo (S)
Gln→Pyro-Glu (N-term Q)	Sulfo (T)
Glu→Pyro-Glu (N-term E)	Sulfo (Y)
Guanidinyl (K)	TMT (K)
ICAT-C (C)	TMT (N-term)
ICAT-C:13C(9) (C)	TMT2plex (K)
ICPL (K)	TMT2plex (N-term)
ICPL:13C(6) (K)	TMT6plex (K)
ICPL:13C(6) (protein N-term)	TMT6plex (N-term)
ICPL:2H(4) (K)	

Source: From http://www.unimod.org.

variable modification. Higher numbers of variable modifications will reduce the specificity and increase the search time. Further modifications can then be taken into account in second-round searches.

Furthermore, free acrylamide in alkylate thiol groups is known to form pro-pionamide, which results in a 71-Da mass shift. An acidic/methanol environment during staining may cause tryptophan oxidation, methylation of aspartic acid, and glutamic acid–rich peptides that may cause unassigned peaks. For the discovery of posttranslational modifications (PTMs) (Chapter 8), another MASCOT algorithm is employed after selecting one or several modifications (Table 1).

The term *peptide tolerance window* refers to the error of experimental peptide mass values. It is strictly dependent on the mass spectrometer employed and the calibration accuracy. The precision gained in the measurements strongly influences the specificity of the search. The peptide mass accuracy should be greater than 100 ppm; otherwise, the number of results will be very high and the probability of correct matching low. It is strongly recommended that the mass accuracy used for database searches be determined experimentally (e.g., by using the digest of a known protein such as bovine serum albumin or by using masses of trypsin autoproteolysis fragments), not just guessed at on the base of the instrument manufacturer's specifications. Filter programs are often used to remove keratins and autolytic peaks from the MS spectra.

Each result has a given score, which is a number representing the identification reliability. The full protein summary report is displayed by selecting the accession number (sequence coverage, nominal mass, pI value, number of matched peptides, mass values error). To ensure that the final result is correct, it is advisable to check that the arginine-containing peptides derived after tryptic digestion are the most abundant peaks on the fingerprint map.

During analysis of high-molecular-mass proteins, one should remember that the results generated will demonstrate low sequence coverage, because compared to low-molecular-mass proteins, the percentage of peptides identified is inferior.

A typical peptide map and peptide mass fingerprinting using a MALDI-TOF instrument are shown in Figure 4.

A major task of database search engines is to differentiate between correct iden-tifications and false-positive hits. The better this differentiation, the easier it is to identify the protein.

Several free proteomic tools are available on the Internet (see Appendix C), such as http://www.expasy.org/toolsr; http://prospector.ucsf.edu; and http://www.unb.br/cbsp/paginiciais/profound.htm.

These tools make possible:

- Theoretical digestion of a chosen protein with a selected enzyme
- Removal of contaminants
- PTM prediction
- Amino acid substitutions
- Prediction of possible oligosaccharide structures from their experimentally de-termined masses

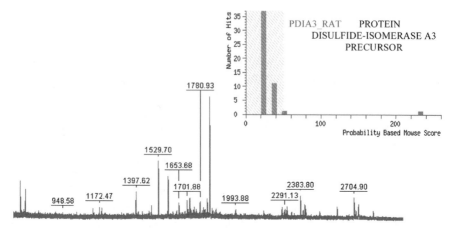

FIGURE 4 Peptide map and PMF identification result.

- Comparison of amino acid composition
- Calculation of the mass of an oligosaccharide structure
- Literature access

PROTEIN IDENTIFICATION USING TANDEM MASS SPECTRA

The peptide sequence tag (PST) technique of protein identification was developed by Mann and Wilm in the mid-1990s. Their discovery led to protein identification using a set of tandem MS/MS spectra. The method also requires protein reduction, alkylation, and digestion, as described earlier in the chapter, but it is more specific and accurate than peptide map analysis (PMF), and is highly sensitive.

Tandem mass spectra contain more structural information then does a peptide map. Short sequences (*sequence tags*) are used for protein identification. The more such tags there are, the more accurate will be identification of a given protein.

The basic idea behind a PST experiment is selection of peptides from the peptide map and identification of their amino acid sequences (Figure 5). An MS/MS spectrum with AA sequence assigned is shown in Figure 6.

- A protein sequence can often be identified based on five or six amino acid residues. We strongly advise identification based on at least four or five sequence tags.
- Careful inspection is necessary to check on even distribution of tags that have been identified along the protein sequence.
- A combination of PMF and sequence tags provides more accurate identification.

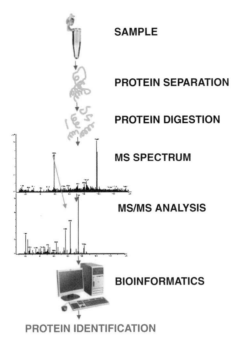

SAMPLE

PROTEIN SEPARATION

PROTEIN DIGESTION

MS SPECTRUM

MS/MS ANALYSIS

BIOINFORMATICS

PROTEIN IDENTIFICATION

FIGURE 5   The PST approach.

- Mass accuracy (and mass difference between calculated and experimental values) is an important criterion.
- The final result depends strongly on the quality of raw data (peak intensities, accuracy, etc.).
- Never trust a printout of proteins identified without further examination of the data. The results must make sense!
- Data need to be verified (e.g., up- or down-regulation) using another technique based on a different principle (e.g., Western blot).

Several algorithms are used to search sequence databases with tandem mass spectra (e.g., PeptideSearch, Sequest, ProteinProspector, Sonar). MASCOT is one examples and is described here in detail.

The MASCOT search engine (Figure 7) matches the experimental fragments to fragments from the database predicted theoretically, starting with the most intense ions. The peptide scores are added to gain the total protein score; thus, the more peptides identified, the higher the confidence in protein identification.

The search result is displayed by MASCOT as shown in Figure 8. The molecular weight search score MASCOT Mowse shows the confidence level of the data obtained.

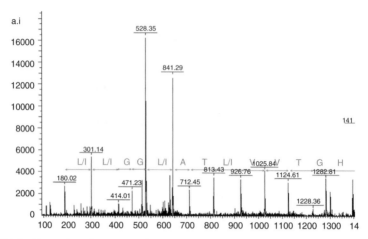

FIGURE 6 Extraction of a sequence based on MS/MS spectra. The mass differences between ions indicate the order of amino acids along the sequence.

MASCOT MS/MS Ions Search

Your name	Email
Search title	
Database	MSDB
Taxonomy	All entries
Enzyme	Trypsin	Allow up to 1 missed cleavages
Fixed modifications	Acetyl (K) / Acetyl (N-term) / Acetyl (Protein N-term) / Amidated (C-term) / Amidated (Protein C-term)	Variable modifications	Acetyl (K) / Acetyl (N-term) / Acetyl (Protein N-term) / Amidated (C-term) / Amidated (Protein C-term)
Quantitation	None
Peptide tol. ±	1.2 Da # 13C 0	MS/MS tol. ± 0.6 Da
Peptide charge	2+	Monoisotopic ● Average ○
Data file	Search...
Data format	Mascot generic	Precursor m/z
Instrument	Default	Error tolerant □
Decoy □		Report top AUTO hits
Start Search ...		Reset Form

FIGURE 7 MASCOT MS/MS ion search interface. Choosing the type of instrument provides additional information regarding the ion series generated by the particular type of instrument. (From http://www.matrixscience.com/cgi/search_form.pl?FORMVER=2&SEARCH=MIS.)

There are several different possibilities for identifing proteins from databases using mass spectral data in a statistical way. Common methods are:

- Probability-based Mowse score (MASCOT)
- Bayesian probability (ProFound, Sonar)
- Cross-correlation (MS-Fit, Sequest)

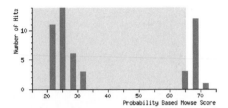

Protein Summary Report

Switch to Peptide Summary Report

To create a bookmark for this report, right click this link: Protein Summary Report (../data/20030107/F024469.dat)

Re-Search All   Search Unmatched

Index

Accession	Mass	Score	Description
1. JC1086	10691	72	myoglobin - giant panda (fragment)
2. MYOT	17026	70	myoglobin - Eurasian river otter
3. MYRB	17079	70	myoglobin - rabbit (tentative sequence)
4. MYHO	16940	70	myoglobin [validated] - horse
5. JN0410	17009	70	myoglobin - European beaver

FIGURE 8   Search results as displayed by MASCOT. The screened area denotes the area where confidence in the result is less than 95%. For a single MS/MS spectrum, this would be a very reasonable result.

Each of these methods has advantages and disadvantages, but the all give a result which states that there is a certain likelihood that the protein name/accession number is the major component of the analyte. A Mowse score below the confidence threshold indicates a result with a reasonable chance of being a false-positive hit (typically, a 5% chance). The absolute score values depend on several factors, such as database size and number of peaks used for the query, as well as number of peaks matched. However, as a very rough guideline, values in the range 50 to 100 suggest that the experiment should be repeated, preferably using another sample or a larger amount of biological material. Score values above 100 can be accepted, still bearing in mind the overall sense of the data produced.

An increase of 10 units in the Mowse score indicates an increase in the confidence in the result of the search by a factor of 10. Therefore, a hit with a Mowse score of 120 is 1000 times more reliable than one with a score of 90. The search engines MASCOT and Profound usually report their results as shown in Figure 9.

Protein Modifications

Identification of protein modifications (PTMs) or mutations is the most difficult task. Peptide maps or peptide tags do not cover the entire protein sequence, and therefore many PTMs are not detected. This fact is particularly important, as PTMs or mutations often influence protein activity and function. PTMs are very common in many proteins, but modification is a statistical process and such modified molecules may not be detected during batch analysis. Generally, enrichment techniques are used

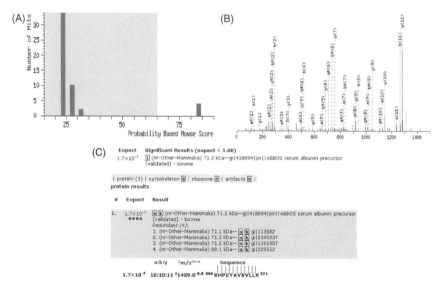

FIGURE 9   Typical presentations of MASCOT and ProFound/Sonar. (A) Typical diagram for MASCOT results. The green shaded area represents the area where the confidence in the result is less than 95%. The red bars show the number of hits (height of the bar) as a function of the Mowse score. (B) Detailed result for an MS/MS spectrum with annotated peaks. (C) Results of a Sonar search using the same data set. Here, an expectation value (it tells us more or less how frequently we may expect a random event that gives a comparable result). The lower this value, the better. There is also an annotation of the fragment ion masses observed as lines above the sequence. (*See insert for color representation of figure.*)

in the search for a particular PTM. For instance, phosphoproteins are concentrated on TiO$_2$ columns, and glycoproteins are enriched with the help of various lectins. More details are provided in Chapter 8. Expertise (and not only computer algorithms) is required for the final analysis of such data.

A useful feature is the automatic detection of some modifications (e.g., phosphorylation, methionine oxidation). For instance, the BioTools software from Bruker Daltonics does this by looking for the presence of peaks in the area of −98 Da and −80 Da (phosphorylation, and neutral loss of H$_3$PO$_4$ and HPO$_3$, respectively) and −64 Da (methionine oxidation, neutral loss of CH$_3$SOH from an oxidized side chain) from the precursor mass. The detection of such a modification is indicated as a small icon in the upper left corner of the sequence window, as shown in Figure 10. This information can be utilized to get an idea whether or not these modifications might be present.

Uninterpreted Mass Spectra

The easiest way to identify a protein is to use uninterpreted MS/MS spectra information for the database search. This is also the most common way at present, since

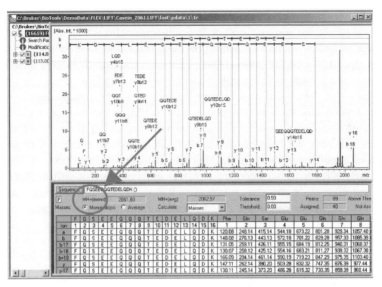

FIGURE 10   Automatic detection and indication of peptide modifications in BioTools. (*See insert for color representation of figure.*)

it is simple and does not require extensive user interaction, which makes it easy to automate. Uninterpreted mass spectra are raw spectra without assignment of ions to any sequence.

The procedure involves detection of the peaks in spectra to create a *peaklist*, and submission of the peaklist into a database search engine (e.g., MASCOT or Sonar). Not all search engines support all types of searches (Sonar and Sequest support only MS/MS searches, and ProFound supports only PMF searches).

New Sequences

In most proteomics experiments, PMF and uninterpreted MS/MS data are sufficient for identification of the protein being analyzed.

> Identification of a given protein based on a few peptide sequences and a search engine does not provide complete information regarding this molecule. It is necessary to reveal all amino acids and all PTMs, mutations, and so on.

It must be considered in the course of protein identifcations that one always identifies peptides based on MS/MS data (in contrast to PMF searches, where proteins are identified). This means that particular attention has to be paid to the *protein inference problem*. A set of identified peptides may indicate the presence of a specific protein or group of proteins (e.g., isoforms). As long as no specific peptides for a particular isoform are detected, it is not possible to conclude that a specific protein isoform is present, only the entire group. As an example, the identification of a peptide originating from the constant area of an immunoglobulin γ chain only indicates the presence of IgG as such; it does not provide information about neither the subtype nor the specificity of the antibody. It might also indicate that 1000 different IgG antibodies are present in the sample, sharing the same peptide.

However, in the case of proteins and peptides of unknown sequences or modified in an unknown way, such an approach is useless. PMF and related strategies are only capable of identifying proteins whose sequences are stored in databases, but many proteins still have not been entered in such databases. An alternative might be the search of genome sequences after computerized translation to protein. The risk is that many DNA sequences contain errors or ambiguities that will be translated to protein sequences. In any case, this approach may provide at least some clues to the protein but should be carefuly verified. Moreover, there is an even higher number of posttranslational modifications that are still unknown, which results in peptides with masses unrecognized by the PMF software. It is possible to define a large number of variable modifications in the database search, but this will increase the statistical threshold for reliable matches significantly, and thus this decision can render the software unable to identify any proteins at all.

In such cases the only possiblity of identifying the protein of interest successfully is to perform de novo sequencing. The term *de novo* means that during the sequencing procedure, no external database is used, just information from the product ion spectrum (i.e., MS/MS data).

DE NOVO SEQUENCING

By far the most demanding but also challenging task when dealing with MS/MS data is a de novo sequencing attempt, the deduction of more or less a complete amino acid sequence from a spectrum without further information. In contradistinction to the methods described earlier, de novo sequencing requires some time and experience to perform, but most manufacturers of mass spectrometry equipment provide software to assist in this task.

Mass spectrometry–based de novo sequencing is the equivalent of chemical sequencing methods such as Edman degradation. The major advantage of the MS-based approach is the ability to sequence peptides of much lower concentration and with modified N-termini. Moreover, de novo sequencing can often be performed successfully in peptide mixtures without prior separation.

In the MS approach it is difficult to distinugish isobaric amino acids, (those having the same or very similar molecular mass), such as Leu and Ile (which have the same elementary composition, 113.084 Da) and Lys/Gln (128.095/128.059 Da; only high-resolution MS can distinguish them). High-energy collisions can be helpful in this case for distinguishing between Leu and Ile. Moreover, it is not uncommon that the MS/MS spectrum quality allows only for determination of the partial sequence of a peptide.

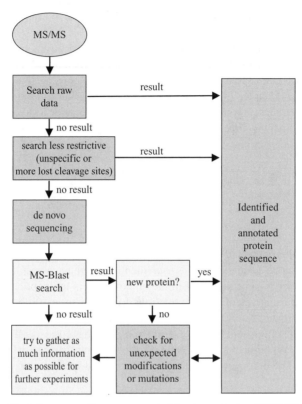

FIGURE 11   Work flow generally recommended for the analysis of MS/MS data. The green boxes are tasks that can be performed using such software as BioTools, while the yellow boxes show tasks that involve other search engines or databases. Generally, this procedure should be followed stepwise from the easiest to the most difficult tasks. A comment to pacify the reader's mind: You will rarely encounter occasions where you will get 100% sequence coverage. (One of the authors (M.M.) has had only one case in eight years that involved a combination of MALDI and ESI as an ionization technique.) Therefore, having gaps or unclear regions in sequences is absolutely normal. This is especially the case when working with unsequenced genomes and having to rely on similar entries in the databases. (*See insert for color representation of figure.*)

There are multiple strategies as to how to perform successful de novo sequencing, but a detailed description is beyond the scope of this chapter. Here we summarize selected basic strategies and guidelines. The general scheme of the procedure is summarized in Figure 11.

We should note that information provided by the product ion spectra depends on both the fragmentation mechanism and the mass analyzer used. Different peptide fragments (ion series) will be visible on the spectrum e.g., obtained from ESI ion-trap instruments with collision-induced (CID) fragmentation, compared with MALDI-TOF/TOF instruments with LID fragmentation. In this chapter we focus primarily on the low-energy CID fragmentation, but except for product ion types, the sequencing procedure is similar to that of other intrument types.

Very rarely can all peaks visible in a product ion spectrum be identified, even in the case of the fragmentation of known peptides. Also, not all peptide fragments provide information usable in sequencing. Some fragments, such as ions from the b and y ion series (Chapter 7.4) provide sequence information and are called *sequence-specific*. Others, such as water or ammonia loss from a peptide's side chain, are called *non-sequence-specific ions*. Calculation rules for ion masses are as follows:

$$b_1 = AA + 1$$
$$b_{n-1} = [M + H]^+ - 18 - AA$$
$$y_1 = AA + 19$$
$$y_{n-1} = [M + H]^+ - AA$$

where AA is an amino acid residue mass.

During de novo sequencing, the user identifies sequence-specific ions, assigns them to the correct ion series, and in the end, discovers the peptide sequence. One should note that masses of consecutive ions from one series differ by the mass of amino acid residues. Annotation of several ions belonging to one series results in the formation of an amino acid ladder, a string of mass shifts corresponding to the amino acid sequence. When such a ladder is found, the sequence can easily be revealed from the spectrum. The first question is: Where does the ladder start, and which ion series does it represent?

Practical Recommendations for De Novo Sequencing

The best practice in de novo sequencing is to start near one end of a peptide, as terminal groups ($-NH_2$ on the N-terminus and $-OH$ on the C-terminus) make it possible to distinguish ions containing N- and C-termini. This can be done either in the low-mass region, where a_1, b_1, and y_1 ions can be detected, or in the high-mass region, where a_{n-1}, b_{n-1}, and y_{n-1} ions can be found (n is the number of amino acids in a peptide).

Identification of low-mass fragments is not possible in all mass spectrometers that utilize MS/MS only. In the commonly used quadrupole ion-trap intruments, a product ion of mass up to 20 to 30% of the precursor mass cannot be detected (mass exclusion by IT). Additional sequencing steps would be necessary (MS^n), but this may not be realistic, in particular when nanochromatography is linked to MS. In other cases the sequencing should start from the low-mass region. The b_1 ion has a mass of amino acid residue (increased by 1 Da for an N-terminal proton), whereas the mass of the a_1 ion is 28 Da lower, due to removal of the CO group. Identification of the first N-terminal amino acid can be achieved by inspecting all possible a_1 and b_1 ions.

A similar procedure can be performed with a y_1 ion whose mass is equal to an amino acid residue of +17 Da for a C-terminal OH group of + 2 Da for two additional protons attracted by ions from these series (see Chapter 7.4 for ion structure).

Inspection of all possible y_1 ion masses often allows for identification of the first C-terminal residue.

Identification of the low-mass region is profitable for yet another reason. During fragmentation at slightly higher energies, double peptide cleavages occur, resulting, among other possibilities, in the formation of immonium ions mass-specific for particular amino acids. Such ions are not sequence-specific themselves, as they do not provide information on the order of AA residues in the sequence, but they are very valuable in confirming that a given amino acid is present in the peptide being analyzed.

If the low-mass region is unavailable or the information obtained is insufficient, one should focus on product ions that have masses slightly lower than that of precursor ions: namely, a_{n-1}, b_{n-1}, and y_{n-1} ions. All these ions have one terminal amino acid detached. In low-energy CID spectra, b_n ions can often be found.

In typical cases either the low- or high-mass region allows for identification of at least one of a peptide's terminal amino acids. Further sequencing should be done by filling the amino acid ladder and finding mass shifts specific to amino acid residues relative to ions identified.

Some operators perform a de novo sequencing procedure in the opposite order, trying to find specific mass shifts first. Such an approach often results in the identification of a partial sequence string, but without knowing whether it originates from the N- or C-termini.

Terminus-specific Fragmentation: T^3

A rather new technique that had been presented as a combination of ISD and PSD for the first time on an ASMS poster in 1997 (D. S. Cornett, "Prompt In-source Fragmentation of Biomolecules in a Gridless Reflectron-TOF") has been reinvented under the term T^3 (Terminus TOF/TOF) by D. Suckau and A. Resemann as a combination of ISD and LID in a TOF/TOF instrument. Figure 12 illustrates the general principle. First, ISD fragment ions originating from the N- or C-terminal region are formed. Next, one of these fragment ions is selected and undergoes metastable PSD fragmentation, the precursor and the fragment are further accelerated in the LIFT unit, and a complete MS/MS spectrum of the fragment ion is obtained.

The appearance of spectra acquired this way is almost identical to that of MS/MS or rather, pseudo-MS^3 spectra. However, depending on the ion type used for isolation and fragmentation, certain modifications might have to be considered:

- Unlike unmodified peptides, a-type fragment ions have a loss of COOH.
- c-type fragment ions behave as peptide amides.
- y-type fragment ions are identical to normal, unmodified peptides.

Taking the foregoing information into account, the analysis is relatively easy. For automated database searches, c- and y-type ions can be searched without any change just by allowing amidation on the C-terminus as a variable modification. Figure 13 shows an example of such a T^3 analysis of β-casein. The ISD spectrum has been acquired in DHB as the matrix. From the ISD spectrum, two precursors had been

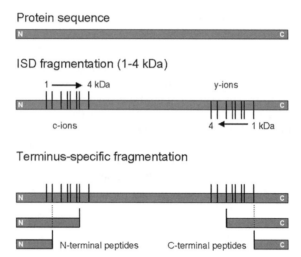

FIGURE 12   Operational principle of T^3 sequencing. Using ISD, N- and C-terminal fragment ions (usually, c and y type) are obtained and undergo further metastable fragmentation, giving the ion series usually observed, which are also obtained by regular LID spectra.

selected for subsequent MS^3 experiments, of which one was a c and one a y ion. A charming effect of this type of search is that it is not necessary to know which of the fragments belongs to which series, since this will be recognized by the search engines automatically, due to the fitting modification. Furthermore, as shown in this example, ISD leaves the labile phosphorylation intact, whereas the post-source decay fragmentation induces neutral loss of the phosphate group as observed in the T3 spectrum of the c_{16} fragment.

Data Verification

Once the sequence has been assigned, it is obligatory to verify it by loading the sequence into a program that calculates all possible fragments. This will be the proof that the sequence has been assigned properly. The fragment ions calculated can then be searched along the spectrum.

In the early 1980s (i.e., at the beginning of biological mass spectrometry), several algorithms were developed for automatic de novo sequencing. Such programs were sometimes effective, but for a limited number of AA residues. Currently available algorithms are far more efficient and are able to reveal sequences from MS/MS data even for large molecules (see also Chapter 7.6).

Using MS-BLAST for the Identification of Proteins

Compared to the use of accurate MS/MS data for database searches, a relatively newer approach is the idea of using the BLAST homologous sequence search tool for protein identification using multiple de novo sequenced peptides [see Shevchenko et al. (2001) for references]. The basic idea is to create multiple sequences which are

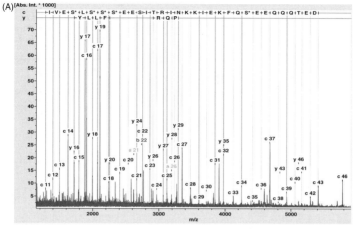

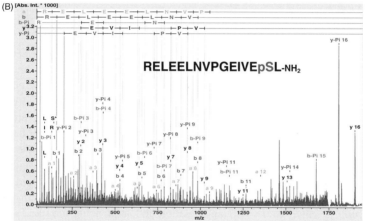

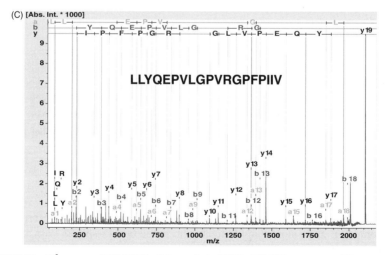

FIGURE 13 T^3 spectra (ISD-LID) of β-casein in DHB: (A) ISD spectrum, T3 spectrum of c_{16} (B) and of y_{19} (C). (*See insert for color representation of figure.*)

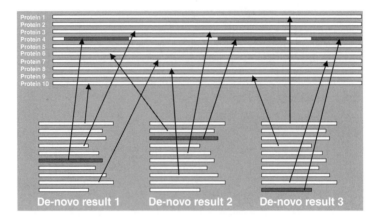

FIGURE 14   Schematic view of the function of MS-BLAST. (*See insert for color representation of figure.*)

not necessarily identical, but more or less similar to the actual peptide sequence for each peptide MS/MS spectrum. The basic principle of MS-BLAST is the following. The software searches for sequence stretches in the database which are homologous to different short sequences entered. Normal BLAST searches would normally need long sequence stretches to get meaningful results but MS-BLAST overcomes this problem by using multiple sequences. For example, we will probably have one hit with weak similarity to each sequence in the sequence set of a peptide. But when we look at three different peptides, the chance that we will find three weakly similar hits, one for each of the three peptides, all fitting one common protein sequence, will be rather small. Figure 14 illustrates the underlying principle of MS-BLAST. A description of other details on fragmentation and sequencing may be found in the recommended reading or in more advanced materials.

Acknowledgments

The kind help of Bruker Daltonics during the preparation of this material is deeply appreciated. This work was supported by International Centre for Genetic Engineering and Biotechnology (ICGEB) grant CRP/POL05-02 and ENACT (European Network for the Identification and Validation of Antigens and Biomarkers in Cancer and Their Application in Clinical Tumor Immunology) contract 503306. Marek Noga was supported by a research grant from the Polish Ministry of Science and Higher Education, grant N204 136 32/3396.

SELF-STUDY QUESTIONS

1. Using the SwissProt database, try to identify human protein after trypsin digestion, DTT, and iodoacetamide treatment, based on MS spectra with m/z 910.46, 1350.81, 1632.87, 1800.93, 1853.96.

2. Compare the results obtained during a search only among human proteins, and then choosing "all entries," based on fingerprinting map (obtained under the same conditions as for the human proteins): 836.40, 874.41, 992.50, 1030.51, 1045.54, 1428.65, 1753.83, 1803.89, 1945.94, 2337.12, 3163.46.

3. Find the length of the signal fragment and the disulfide bridges in the protein sequence acquired in question 2.

4. Why may sequential MS (MS^n) be difficult to perform when nano-LC-MS linked to mass spectrometry is used?

RECOMMENDED READING

Domon B., Abersold R. Mass spectrometry and protein analysis. *Science* 312(2006) 212–217.

Garbis S., Lubec G., Fountoulakis M. Limitations of current proteomics technologies. *J. Chromatogr. A* 1077(2005) 1–18.

Henzel W.J., Watanabe C., Stults J.T. Protein identification: the origins of peptide mass fingerprinting. *J. Am. Soc. Mass Spectrom.* 14(2003) 931–942. http://www.ionsource.com/tutorial/DeNovo/introduction.htm.

Mann M., Hojrup P., Roepstorff P. Use of spectrometric molecular weight information to identify proteins in sequence databases. *Biol. Mass Spectrom.* 22(1993) 338–345.

Pappin D.J., Hojrup P., Bleasby A.J. Rapid identification of proteins by peptide-mass fingerprinting. *Curr. Biol.* 6(1993) 327–332.

Resing K.A., Ahn N.G. Proteomics strategies for protein identification. *FEBS Lett.* 579(2005) 885–889.

Shevchenko A., Sunyaev S., Loboda A., et al. Charting the proteomes of organisms with unsequenced genomes by MALDI-quadrupole time-of-flight mass spectrometry and BLAST homology searching. *Anal. Chem.* 73(2001) 1917–1926.

Suckau D., Resemann A. T^3 sequencing: Targeted characterization of the N- and C-termini of undigested proteins by mass spectrometry. *Anal. Chem.* 75(2003) 5817–5824.

Thiede B., Hohenwarter W., Krah A., *et al.* Peptide mass fingerprinting. *Methods* 35(2005) 237–247.

7.6

CHEMICAL DERIVATIZATION FOR SEQUENCE ANALYSIS

Adam Moszczynski, Ulf Hellman, Tomasz Dylag, Marcus Macht, and Jerzy Silberring

Chemically assisted fragmentation serves for enhanced fragmentation of peptides and proteins, and also for more reliable de novo sequencing. It is used in MALDI-mass spectrometry, where complicated post-source decay (PSD) mechanisms usually obscure the analysis of fragmentation spectra. The term has been restricted to one specific reagent (CAF from Amersham Pharmacia) but has a much broader meaning. Various reagents are used and can be divided into two groups: chemicals for derivatization, and matrices.

Derivatization is a procedure that utilizes chemical reactions with covalent bond formation. The aims of derivatization are:

- To promote charge site–initiated fragmentation
- To enhance detection of one ion series
- To improve fragmentation yield
- To simplify data interpretation (for de novo sequencing)

Here we focus on a few reactions that have the most important impact on proteomics and de novo sequencing procedures.

CAF: 2-SULFOPROPIONIC ACID NHS ESTER

In the CAF method, the N-terminal amino group of each tryptic peptides is modified by introduction of a sulfonic acid moiety that carries a negative charge. Therefore,

Proteomics: Introduction to Methods and Applications, Edited by Agnieszka Kraj and Jerzy Silberring
Copyright © 2008 John Wiley & Sons, Inc.

during PSD measurement, only peptide bonds are fragmented to produce exclusively b- and y-type ions, but neither a and c, nor x and z, ions. Due to the presence of a negative charge at the N terminus, b-type fragments have zero net charge and are not detected; therefore, in the spectra, only y-type ion series are seen that are easy to interpret.

Sulfonation of the peptide N-terminal residue requires that basic residues are present on its C-terminus. This is perfectly suited with tryptic peptides that contain C-terminal arginine or lysine residues. In order to preserve the ε-NH₂ lysine moieties, they are protected against sulfonation by guanidination with O-methylisourea hydrogen sulfate, which increases peptide mass by 42 Da.

The following stages of CAF derivatization are shown in Figures 1 and 2, respectively. The MALDI mass spectrum obtained after derivatization is searched for peptides with modified Arg and/or Lys residues that can be recognized by mass

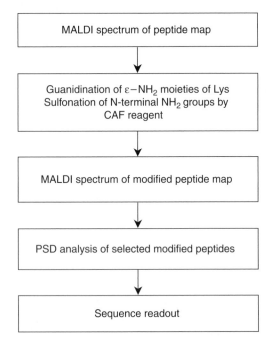

NHS ester of 3-sulfopropionic acid   CAF derivative [mass addition +136 Da]

FIGURE 1   Derivatization with CAF reagent.

```
        ┌─────────────────────────────────────┐
        │   MALDI spectrum of peptide map     │
        └─────────────────────────────────────┘
                        │
                        ▼
        ┌─────────────────────────────────────┐
        │  Guanidination of ε−NH₂ moieties of Lys │
        │  Sulfonation of N-terminal NH₂ groups by │
        │             CAF reagent             │
        └─────────────────────────────────────┘
                        │
                        ▼
        ┌─────────────────────────────────────┐
        │  MALDI spectrum of modified peptide map │
        └─────────────────────────────────────┘
                        │
                        ▼
        ┌─────────────────────────────────────┐
        │  PSD analysis of selected modified peptides │
        └─────────────────────────────────────┘
                        │
                        ▼
        ┌─────────────────────────────────────┐
        │         Sequence readout            │
        └─────────────────────────────────────┘
```

FIGURE 2   Peptide sequencing scheme by CAF derivatization.

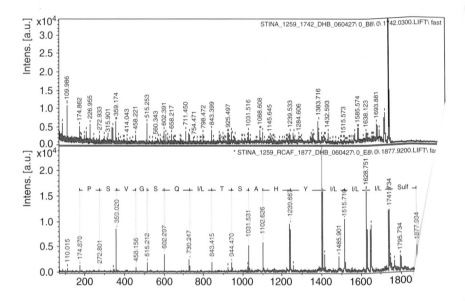

FIGURE 3 PSD of a tryptic peptide from an Ig light chain. Upper panel: native lower panel: after sulfonation.

increase of 136 or 178 Da, (42 Da increase for guanidination of amino Lys residues), respectively, compared to the original peptide map. For t peptides, PSD spectra are registered. They show series of peaks correspo amino acid sequence, with consecutive residues falling off the N-termin presents a difference in a quality of fragmentation pattern for a native after sulfonation with CAF.

Spectra with and without sulfonation are also shown in Figure 4 digest from a *Trypanosoma cruzi* protein. The difference in quality o not as clear at first glance, but further fragmentation of several ions (F gives a clear answer as to how useful the technique is.

SPITC: 4-SULFOPHENYLISOTHIOCYANATE

Sometime in 2008, CAF will probably be replaced by SPICT. It is portant to mention this alternative methodology, which has been u and is simpler and cheaper. SPITC also generates sulfonation of p terminus, but in contrast to CAF, it does not bind to the C-termin: always present when trypsin is used in the preparation of pept blocking the lysine ε-NH$_2$ moieties by guanidination is not rec

SPITC selectively enhances detection of the y-ion series in t analysis.

FI
287

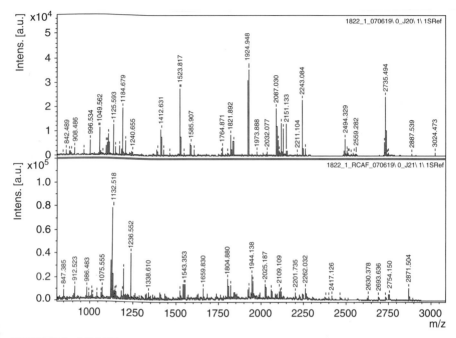

FIGURE 4   MALDI analysis of a trypsin digest from a *Trypanosoma cruzi* protein. Upper panel: spectrum of the native digest; lower panel: spectrum after sulfonation.

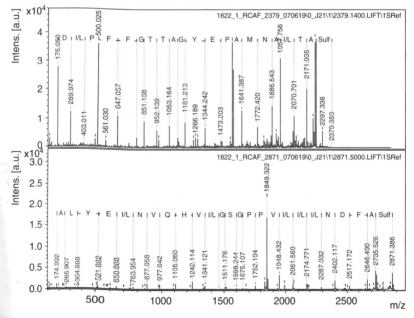

URE 5   PSD of two sulfonated peptides from Figure 4: *m/z* 2379 (upper panel) and *m/z* (lower panel). Twenty-five AA residues were revealed from the latter, low-abundant ion.

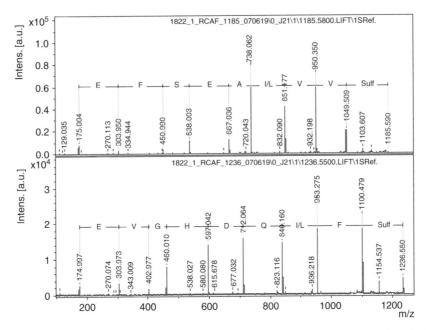

FIGURE 6   PSD of two sulfonated peptides from Figure 4: *m/z* 1185 (upper panel) and *m/z* 1236 (lower panel).

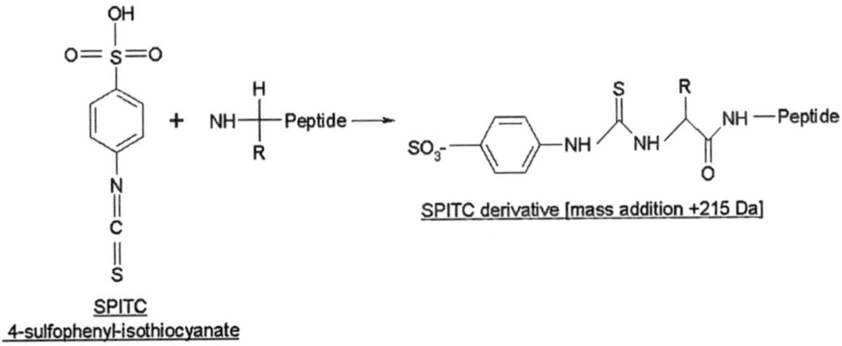

FIGURE 7   Derivatization with SPICT reagent.

ACETYLATION AND DEUTEROACETYLATION

Acetylation and deuteroacetylation utilizes a mixture of acetic anhydride and deuterated acetic anhydride (1 : 1 vl/vl) in methanolic solution and, if performed properly, labels an N-terminal amino group only (Figure 8). The molecular mass of deuterated reagent is higher than standard acetic anhydride by 3 Da; therefore, the labeled peptide being analyzed in the mass spectrometer is represented by two pseudomolecular

$$(CH_3CO)_2O \qquad\qquad (CD_3CO)_2O$$

FIGURE 8   Acetic anhydride (left) and deuterated acetic anhydride (right). Note three deuterium atoms causing a mass shift of 3 Da.

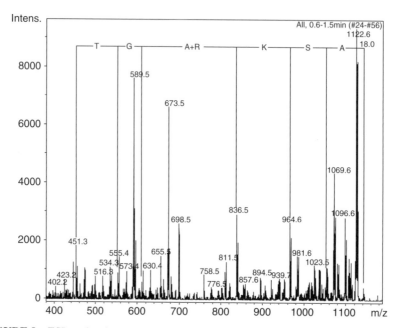

FIGURE 9   ESI product ion spectrum (positive-ion mode) of nociceptin/orphanin FQ (1–11) fragment (FGGFTGARKSA), $[M+H]^+ = 1140.6$ Th, after derivatization with a mixture of acetic anhydride/deuterated acetic anhydride (1:1 vl/vl). Annotated peaks belong to labeled b-ion series, allowing easy determination of the C-terminal part of a sequence. Nonlabeled doublet peaks belong to a-ion series or result from side-chain fragmentation.

ions forming a characteristic "double peak" (two peaks separated by 3 Th; Figure 6). In contrast to CAF or SPITC, this reaction does not promote fragmentation but serves for easier identification of fragment ions (de novo sequencing; see Chapter 7.4).

The major feature of this labeling technique is that all N-terminal ions formed during fragmentation are labeled, whereas C-terminal fragments are not labeled at all. As a result, a characteristic spectrum with labeled ions is observed and the sequence can easily be revealed by calculating the distances between peaks. An example is shown in Figure 9. Such a procedure can be fully automated.

^{18}O LABELING

A relatively easy way to differentiate between fragment ions originating from the N-terminus and C-terminus of a peptide is to introduce a stable isotope label in

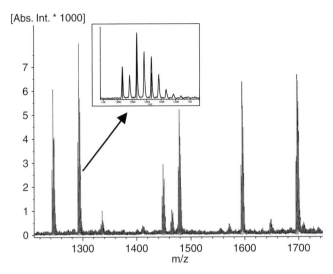

FIGURE 10   MALDI mass spectrum of a proteolytic peptide mixture after enzymatic diges-tion in a mixture of ^{16}O and ^{18}O water. The inset shows a zoomed view to illustrate the isotopic peak pattern with the signal pairs resulting from incorporation of the isotopic mixture. (*See insert for color representation of figure.*)

the form of ^{18}O. This can be done during digestion of an intact protein. When a proteolytic enzyme cleaves a protein into peptides, one H_2O molecule is added to the newly formed peptide termini. This water is taken from the solvent. When the solvent consists of a mixture of water containing ^{16}O and ^{18}O, this results in a statistical incorporation of both atom types into the peptide. Thus, every peptide appears as a pair of peaks separated by 2 Th, as shown in Figure 10.

Upon fragmentation of these peptides, the resulting MS/MS spectrum will show two types of signals: single signals and signal pairs. The first class consists of fragment ions that do not contain the C-terminus (usually, i-, a-, or b-type ions), and the second consists of fragment ions containing the C-terminus (usually, y-type ions). Figure 11A shows a typical spectrum obtained from an isotopically labeled peptide, while in Figure 11B the y_1 ion of an arginine-terminated tryptic peptide is shown, clearly demonstrating incorporation of the $^{16}O/^{18}O$ mixture and showing the typical isotopic pattern.

DAN: 1,5-DIAMINONAPHTALENE

DAN (Figure 12) serves as a matrix in the MALDI method. Primarily, the reagent served as a matrix for MALDI-TOF analysis of gangliosides, but its reducing properties were also observed (characterizing disulfide bridges). Moreover, this matrix contributes to pronounced in-source fragmentation [in-source decay (ISD)]

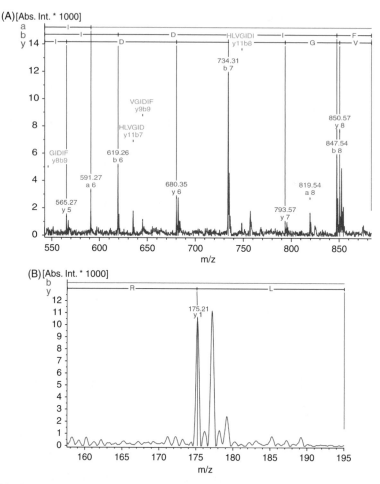

FIGURE 11   (A) MALDI-MS/MS spectrum of a tryptic peptide. The annotated pairs always belong to y-type ions, while singlet peaks are either b-type, a-type, or internal fragment ions. (B) Zoomed view of the y^1 ion of a tryptic peptide with an arginine at the C-terminus. The doublet structure with 2 Da separation for the ^{16}O and the ^{18}O species is clearly visible. There is also a small third peak resulting from double incorporation of ^{18}O (from reentering of an already cleaved peptide into the active site of the enzyme and a second oxygen exchange by the enzyme). The longer the incubation time, the more pronounced the effect. It cannot be avoided completely. (*See insert for color representation of figure.*)

of peptides and proteins during MALDI analysis. 1,5-DAN and its derivatives are highly suitable for top-down proteomics [i.e., identification of intact proteins without preparation of peptide maps (the bottom-up approach)]. Figure 13 shows the spectrum of heptadecapeptide dynorphin A and its revealed sequence (60% coverage). As expected, the presence of Pro in the sequence obscures formation

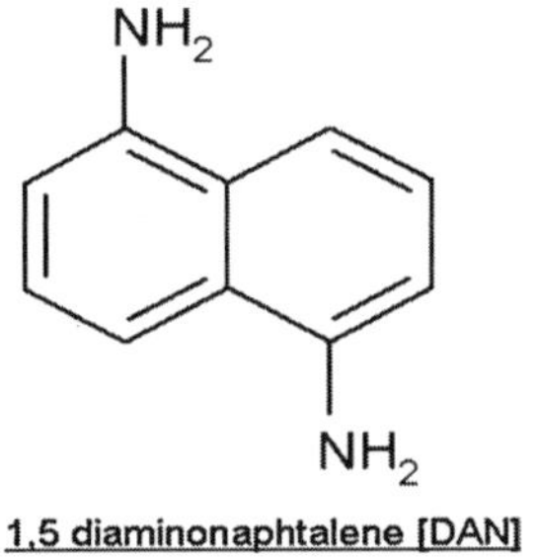

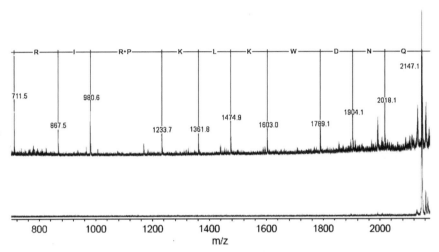

1,5 diaminonaphtalene [DAN]

FIGURE 12   1,5-Diaminonaphtalene.

FIGURE 13   Analysis of neuropeptide dynorphin A (YGGFLRRIRPKLKWDNQ) by MALDI- TOF. Upper panel: in 1,5-DAN; bottom panel: in 1,5-DAN : CHCA (1 : 1 w/w).

of the c ion. Other abundant peaks present in the spectrum do not belong to the dynorphin A sequence and arise from impurities present in the sample.

The method is easy to carry out, and sequence coverage often reaches 70 to 80% for smaller proteins. A C-ion series is usually most abundant along the spectrum. This method can be used as a complementary technique together with other methods, such as TOF/TOF fragmentation, and is advantageous over the post-source decay (PSD) fragmentation. Another important feature is that ISD can be used for high-molecular-mass compounds up to at least 100 kDa, in contrast to PSD, which is suitable for molecules up to 3 to 5 kDa.

A similar approach utilizes ferulic acid, but this matrix has no reducing properties and reveals about 7% of the sequence.

Acknowledgments

The kind help of Bruker Daltonics during the preparation of this material is deeply appreciated. This work was supported by International Centre for Genetic Engineering and Biotechnology (ICGEB) grant CRP/POL05-02. Tomasz Dylag received a scholarship from the Foundation for Polish Science.

RECOMMENDED READING

Connolly J.B., Larsson H., Openshaw M., Barnes A. Improved confidence in protein identification using MASCOT via peptide SPITC derivatization. *Shimadzu-Biotech Application Note 202-ASMS*, 2005.

Chen P., Nie S., Mi W., Wang X.C., Liang S.P. De novo sequencing of tryptic peptides sulfonated by 4-sulfophenyl isothiocyanate for unambiguous protein identification using post-source decay matrix-assisted laser desorption/ionization mass spectrometry. *Rapid Commun. Mass Spectrom.* 18(2004) 191–198.

Flensburg J., Tangen A., Prieto M., Hellman U., Wadensten H. Chemically-assisted fragmentation combined with multi-dimensional liquid chromatography and matrix-assisted laser desorption/ionization post source decay, matrix-assisted laser desorption/ionization tandem time-of-flight or matrix-assisted laser desorption/ionization tandem mass spectrometry for improved sequencing of tryptic peptides. *Eur. J. Mass Spectrom.* 11(2005) 169–179.

Keough T., Lacey M.P., Youngquist R.S. Derivatization procedures to facilitate de novo sequencing of lysine-terminated tryptic peptides using postsource decay matrix-assisted laser desorption/ionization mass spectrometry. *Rapid Commun. Mass Spectrom.* 14(2000) 2348–2356.

Noga M.J., Lewandowski J.J., Suder P., Silberring J. An enhanced method for peptides sequencing by N-terminal derivatization and mass spectrometry. *Proteomics* 5(2005) 4367–4375.

Noga M.J, Asperger A., Silberring J. N-Terminal H(3)/D(3)-acetylation for improved high-throughput peptide sequencing by matrix-assisted laser desorption/ionization mass spectrometry with a time-of-flight/time-of-flight analyzer. *Rapid Commun. Mass. Spectrom.* 20(2006) 1823–1827.

Quinton L., Demeure K., Dobson R., Gilles N., Gabelica V., De Pauw E. New method for characterizing highly disulfide-bridged peptides in complex mixtures: application to toxin identification from crude venoms. *J. Proteome Res.* 8(2007) 3216–3223.

8

PROTEIN MODIFICATIONS

SIMONE KÖNIG

Although the code for the formation of proteins is laid down in the genetic information, the protein product may deviate from the programmed sequence due to functionally directed processes and/or environmental influences. As Figure 1 shows, on the DNA level, coding errors or mutations may cause changes in the protein sequence. For instance, a single amino acid change from hydrophilic glutamic acid to hydrophobic valine in hemoglobin α is responsible for the formation of sickle-shaped red blood cells. This placement of a nonpolar residue on the outside of hemoglobin causes disadvantageous structural changes in red blood cells but is also correlated with the protection against malaria.

The elucidation of protein modifications is a necessity in functional investigations. Mass spectrometry is a prime technique in detecting structural changes, due to its ability to measure molecular mass. The E $\rightarrow$ V mutation mentioned above causes a 30-Da shift to lower mass compared to regular hemoglobin α. As shown in Figure 2, intact human hemoglobin can easily be measured that detect molecular-mass differences with errors better than 1 Da. In the example a triple-quadrupole mass spectrometer was used to run hemoglobin as extracted from a drop of blood. Two major charge-state envelopes were observed for α and β chains which can be deconvoluted to give masses of 15,128 and 15,869 Da, respectively. Comparison with the values calculated from the respective SwissProt entries (P69905 and P68871) leads to the conclusion that the N-terminal methionine has been removed in both chains.

Proteomics: Introduction to Methods and Applications, Edited by Agnieszka Kraj and Jerzy Silberring
Copyright © 2008 John Wiley & Sons, Inc.

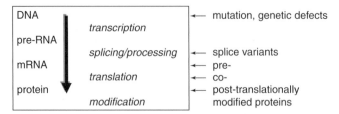

FIGURE 1 Transcription and translation of the genetic code. Deviations from expected protein sequences can be due to changes on the gene level, splice variants, or structurally modified amino acid residues.

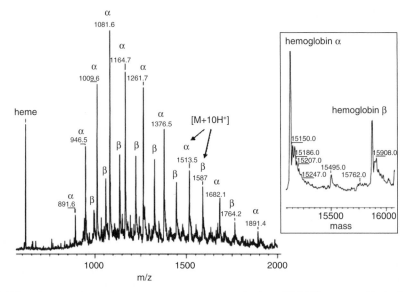

FIGURE 2 Blood protein extract as measured on a Finnigan TSQ700. Detected is the heme molecule liberated from the globin. Hemoglobin chains α and β are observed as charge-state envelopes carrying 8 to 17 protons which can be deconvoluted for molecular-mass calculation as shown in the box on the right. This sample is owned and was measured by the author at the laboratory of H. Fales, National Institutes of Health, Bethesda, Maryland.

CLEAVAGE

With that, one major cotranslational protein modification is introduced: the hydrolytic clipping of N-terminal methionine by methionine aminopeptidases. Methionine is encoded by a single codon (AUG) and carries the start signal for the ribosome to begin protein translation from mRNA. Therefore, it is the first amino acid residue in proteins of eukaryotes and archaea, and most sequences found in public databases still contain it, a fact that must be observed in analysis. Met cleavage is often accompanied by cotranslational irreversible enzymatic N-terminal acetylation.

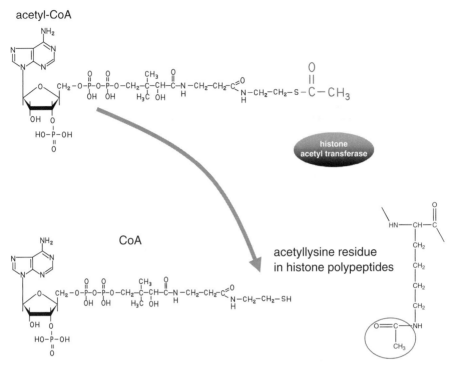

FIGURE 3   Enzymatic lysine acetylation in histones. CoA acts as the donor for the acetyl moiety. (Modified from a design by ScienceSlides. Used with permission from ViviScience Corp., Chapel Hill, NC.) (*See insert for color representation of figure.*)

N-TERMINAL MODIFICATION

N-Terminal acetylation occurs in about 50% of yeast proteins and up to 90% of higher eukaryotic proteins. Acetyl-coenzyme A (CoA) acts as the donor substrate to nucleophilic side chains in proteins, not only at the N-terminus. The best known example is the lysine ε-NH_2 acetylation in histones, which influences the chromatin structure (Figure 3). Histones can also be methylated. N-, O-, and S-methylation are known in nature. For an overview on acylation, see Creasy (2004).

LIPID ANCHORS

Long-chain CoA is also known to act as a donor of myristoyl and palmitoyl chains. These C_{14} and C_{16} acyl chains are attached as lipid anchors to direct the modified proteins to cellular membranes. N-Terminal glycine is often cotranslationally myristoylated catalyzed by N-myristoyl transferase. An example is the Nef protein of human immunodeficiency virus type 1, which binds its own myristoylated N-

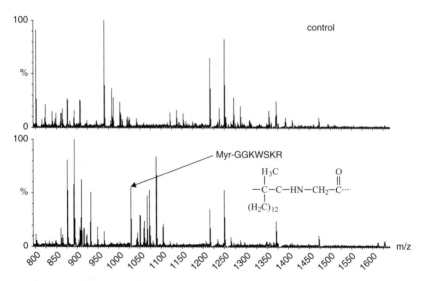

FIGURE 4 MALDI tryptic peptide maps of Nef protein and its myristoylated form (bottom spectrum) measured on a TofSpec-2E MALDI-TOF instrument (Micromass/Waters). Peptide Myr-GGKWSKR is visible as a distinct peak. The spectra were run in collaboration with the group of D. Willbold, Research Center Jülich, Germany.

terminus. Proteomic mass spectrometry is able to detect this modification following tryptic digest of the control and treated protein (Figure 4). This is often the first step in looking for a modified site. In many cases, however, the respective peptide fails to be detected, and purification and analysis protocols optimized for the special chemistry of a particular modification have to be developed to access the modification. This is particularly true for phosphorylation, as discussed below.

C-TERMINAL MODIFICATION

The C-terminus of proteins that participate in signal transduction and protein targeting may be modified by farnesylation (C_{15}; Figure 5) or geranylgeranylation (C_{20}). The prenyl groups are attached to cysteine residues in the C-terminal sequence CAAX by thioether linkages. Such hydrophobic C-terminus allows insertion into the plasma membrane, as is the case for the ras protein. Unfarnesylated ras is not able to transduce growth signals.

Above, a number of important processes have been introduced which lead to modified proteins: namely, cleavage, acylation, and N- and C-terminal modification. Table 1 was generated in an effort to group protein modifications and name some examples. More than 300 modifications of amino acid residues are known. They occur before, during, and after translation, but often they are only referred to as posttranslational modifications (PTMs). Any of the common amino acid residues can be affected, although some are more reactive than others.

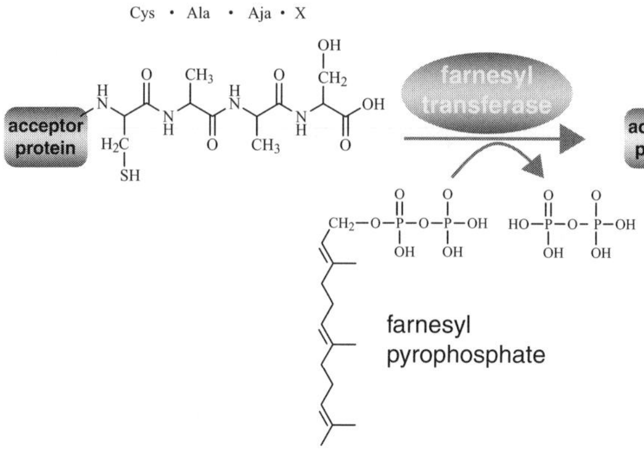

FIGURE 5 Farnesylation at the C-terminus with the activated donor farnesyl prophosphate. The reaction occurs at Cys Ala Ala X sequences. (Modified from a design by ScienceSlides. Used with permission from ViviScience Corp., Chapel Hill, NC.) (*See insert for color representation of figure.*)

UBIQUITINATION

Of the more complex modifications, ubiquitination may be mentioned. Ubiquitin is a highly conserved 8-kDa protein that tags proteins for destruction. Thereby denatured, damaged, or improperly translated proteins are removed from cells by a complex structure, the proteasome. The C-terminal glycine of ubiquitin is bound to the ϵ-NH$_2$ group of lysine via a pathway depicted in Figure 6. Due to the size of the attached moiety, classical peptide mapping approaches provide little information in ubiquitination analysis, and the current research involves complex experiments using cross-linkers.

ADP RIBOSYLATION

ADP ribosylation is an additional example of the attachment of complex structures to amino acid residues. It has been described on arginine, asparagines, cysteine, serine, and glutamic acid (Figure 7). Poly(ADP ribosyl) polymerase (PARP) catalyzes the formation of polymer of ADP-ribose from nicotinamide adenine dinucleotide. One important function of PARP is in the repair of single-stranded DNA nicks. For the analysis of such protein modification, attention has to be paid to the different chemical properties of the phosphate and ribose molecules, which is also true in studies of phosphorylation (see below) and glycosylation.

GLYCOSYLATION

The latter concerns the co- and posttranslational addition of saccharides (Table 2) to membrane and secreted proteins and lipids. The majority of proteins synthesized in

TABLE 1   Examples of Protein Modifications with Functional Relevance in Biological Systems

Cleavage:
- Co- or posttranslational cleavage of initiator methionine in most cytoplasmic proteins
- Cotranslational cleavage of signal peptides during translocation across endoplasmatic reticulum membrane for proteins secreted
- Maturation of proproteins (e.g., formation of peptide hormones from proopiomelanocortin)

End-group modification:
- Acetylation of N-terminal amino acid of cytoplasmic proteins seems to be related to protein turnover
- Acylation of N-terminal amino acid residue for protein interaction with cell membrane or its receptors
- Farnesylation or geranylgeranylation on the C-terminus in signal transduction and protein targeting

Tertiary structure formation:
- Disulfide bond formation (e.g., in insulin, growth factors)
- 4-Hydroxylproline formation in collagen fiber stabilization
- Transpeptidation between γ-carboxyl groups of glutamine residues and free amino groups of lysines by fibrin-stabilizing factor

Reversible:
- Phosphorylation of serine, threonine, tyrosine, and histidine is of importance in regulation and signal transduction
- Acetylation of lysine residues, methylation, ADP ribosylation, and phosphorylation of histones modulate charge and hydrogen binding and are important in chromatin formation
- Acylation and deacylation of chymotrypsin in catalysis

Special function examples:
- Iodination of thyroglobulin tyrosine in thyroid gland hormone action
- Ubiquitinylation tags proteins for destruction
- Increase in hydrophilicity or hydrophobicity by addition of carbohydrate units to asparagine or fatty acids to an α-amino group or cystein sulfhydryl group
- Insufficient carboxylation of glutamate (γ-carboxyglutamate) in prothrombin can lead to hemorrhage on vitamin K deficiency

the endoplasmatic reticulum are glycosylated (Figure 8). The process is catalyzed by enzymes and is site specific, as opposed to the nonenzymatic chemical reaction of glycation. N- and O-Glycosylation are distinguished. For N-linked sugars the consensus sequence Asn-Xxx-Ser/Thr/Cys is known; O-linkage is formed at the side chains of serine or threonine. Less commonly, oligosaccharides may also be linked via the hydroxyl group of hydroxylysine, hydroxyproline, and tyrosine. Typical core structures are depicted in Figure 9. O-linked glycans are not so well defined. Many functions are associated with glycosylation, best known is probably the fact that carbohydrates attached to glycoproteins specify four types of human blood. Glycoproteins exhibit micro- and macroheterogeneity, due to the fact that the length and position of the oligosaccharide chain can vary within a protein. This is one of the reasons why blood and serum proteomic analysis is such a difficult task.

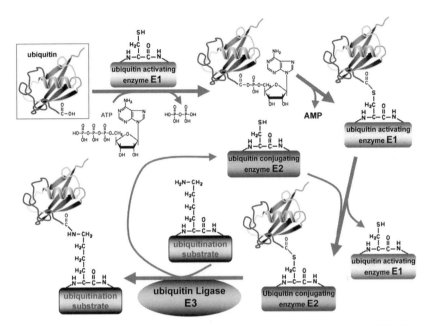

FIGURE 6   Ubiquination on lysine of proteins destined to be degraded. (Modified from a design by ScienceSlides. Used with permission from ViviScience Corp., Chapel Hill, NC.) (*See insert for color representation of figure.*)

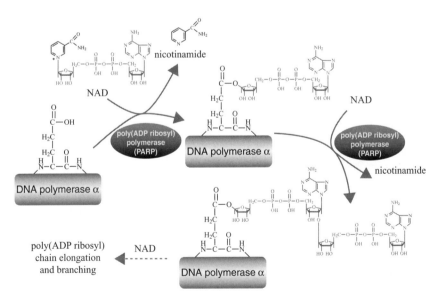

FIGURE 7   ADP ribosylation of glutamic acid catalyzed by PARP. (Modified from a design by ScienceSlides. Used with permission from ViviScience Corp., Chapel Hill, NC.) (*See insert for color representation of figure.*)

TABLE 2 Sugar Moieties in Glycoconjugates

Moiety	Abbreviation	Composition	Mono-isotopic Molecular Mass	Residue Molecular Mass	Low-Mass CID Marker Ions
Galactose	Gal (Hex)	$C_6O_6H_{12}$	180.0634	162.0528	163
Mannose	Man (Hex)	$C_6O_6H_{12}$	180.0634	162.0528	163
Fucose	Fuc (dHex)	$C_6O_5H_{12}$	164.0685	146.0579	147
Sialic acid	NANA	$C_{11}O_9NH_{19}$	309.1060	291.0954	292 274 256
N-Acetylgalactoseamine	GalNAc (HexNAc)	$C_8O_6NH_{15}$	221.0899	203.0794	204
N-Acetylglucoseamine	GlcNAc (HexNAc)	$C_8O_6NH_{15}$	221.0899	203.0794	204
Hex-HexNAc		$C_{14}O_{11}NH_{25}$	383.1428	365.1322	366
Hex$_3$HexNAc$_2$		$C_{46}O_{35}N_2H_{76}$	910.3278	892.3172	

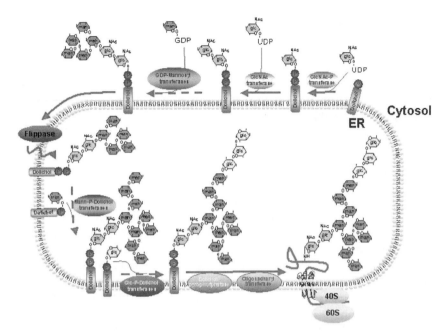

FIGURE 8 N-Glycosylation pathway in the endoplasmatic reticulum. (Modified from a design by ScienceSlides. Used with permission from ViviScience Corp., Chapel Hill, NC.) (*See insert for color representation of figure.*)

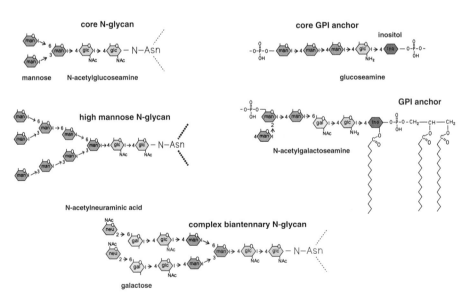

FIGURE 9 Protein glycan modifications. (Modified from a design by ScienceSlides. Used with permission from ViviScience Corp., Chapel Hill, NC.) (*See insert for color representation of figure.*)

In general, the glycan and peptide parts of the glycoprotein need to be analyzed separately. The release of N-glycans can be achieved using peptide-N^4-(N-acetyl-β-glucosaminyl)asparagine amidase F (PNGase F). O-glycans are cleaved on reductive β-elimination. For an overview of carbohydrate and glycoconjugate chemistry, see the book by Robyt (1998).

PHOSPHORYLATION

Protein phosphorylation is a critical process in many biological pathways, such as signal transduction. The phosphorylation studied best so far is the reversible reaction on serine, threonine, and tyrosine catalyzed by enzymes (Figure 10). It serves here as an example to illustrate some analytical possibilities that mass spectrometry offers to tackle modifications. Phosphogroups are difficult to detect in general, and this problem is multiplied for acid-labile bonds. Acid-labile phosphorylation on phosphoamidates of histidine, arginine, and lysine, and acylphosphates of aspartic and glutamic acid, and S-phosphorylation on cysteine (Figure 11) have therefore hardly been investigated. Nevertheless, the phosphoprotein is stable to some extent in standard acidic electrospray solvents, allowing measurement of the

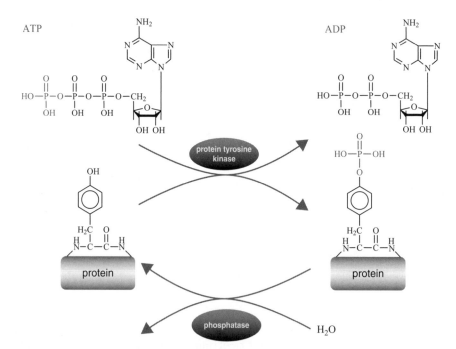

FIGURE 10   Enzyme-catalyzed phosphorylation of tyrosine residues using ATP as the donor substrate. (Modified from a design by ScienceSlides. Used with permission from ViviScience Corp., Chapel Hill, NC.) (*See insert for color representation of figure.*)

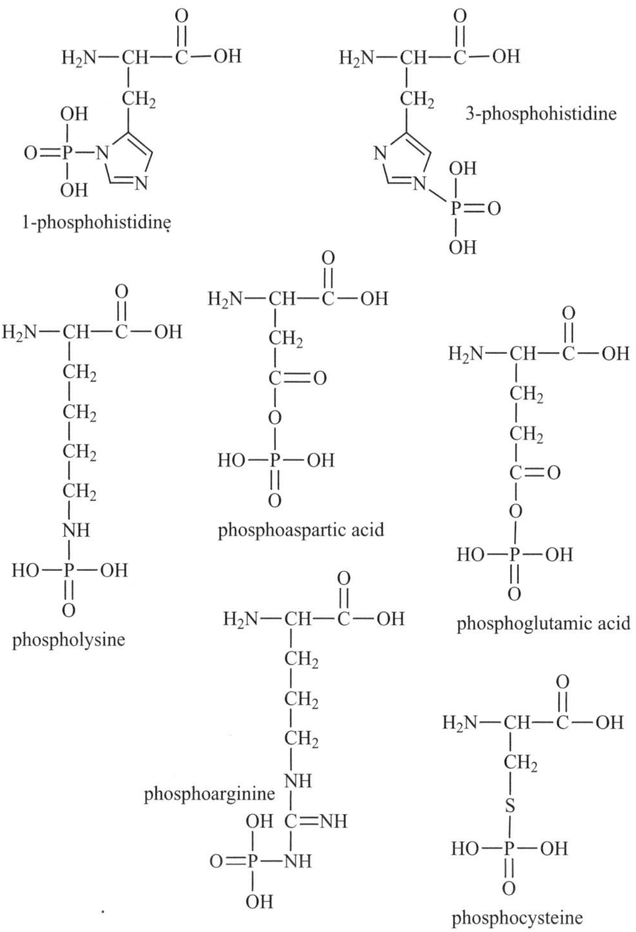

FIGURE 11 Amino acids carrying acid-labile phosphate groups.

intact protein, as shown in Figure 12 for His-phosphorylated nucleoside diphosphate kinase. MALDI-TOF-MS visualizes the mass difference between phosphorylated and unphosphorylated protein in a more obvious manner (Figure 13). In this case, alkaline phosphatase was used as a dephosphorylating enzyme, showing that it is a useful tool to determine the number of phosphorylation sites. To access the sites of modification, enzymatic digest of the protein is a standard procedure accompanied by phosphopeptide-specific purification procedures. One method that is used often is immobilized-metal ion chromatography (IMAC), which enriches phosphopeptides via a chelate complex (Figure 14). Unfortunately, this method also unspecifically purifies other acidic peptides. Nevertheless, in particular in combination with enzymatic dephosphorylation, the method may lead to the assignment of phosphorylation sites, as shown in Figure 15 for myeloid-related protein.

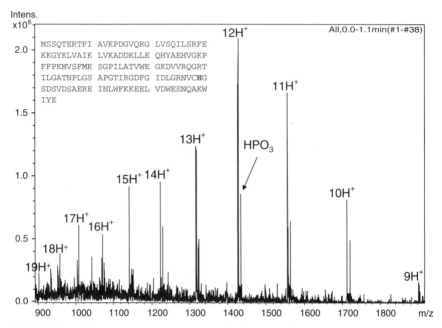

FIGURE 12 Electrospray mass spectrum of intact nucleoside diphosphate kinase phosphorylated on His119 (bold). The protein is measured as a charge-state envelope with an ion trap (Esquire$_{3000}$, Bruker Daltonics). The phosphorylated form can be recognized by a 80-Da mass difference manifested in a second charge-state envelope of satellite peaks. The sample was measured in collaboration with A. Hasche and S. Klumpp of the Institute of Pharmacology and Medical Chemistry, Münster, Germany.

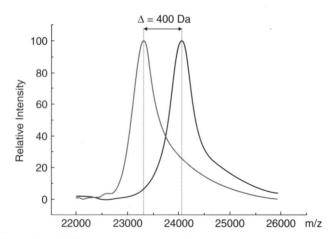

FIGURE 13 MALDI-TOF spectrum of β-casein. This protein carries five O-phosphorylation sites. After removal of the phosphate by alkaline phophatase, the protein mass is decreased by 400 Da. (Modified from a schematic generated by M. Zeller during his Ph.D. work, Münster, Germany, 2004) (*See insert for color representation of figure.*)

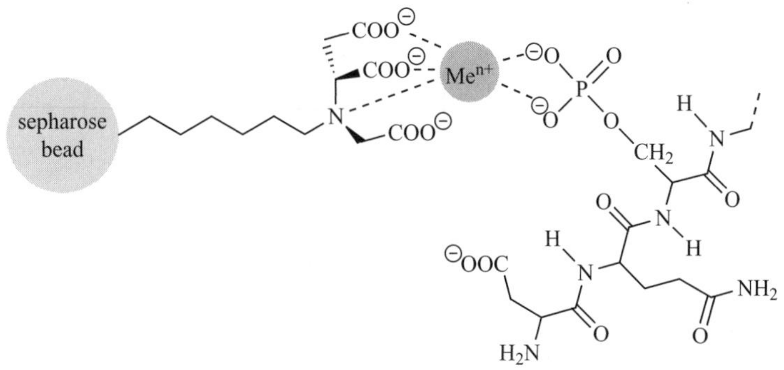

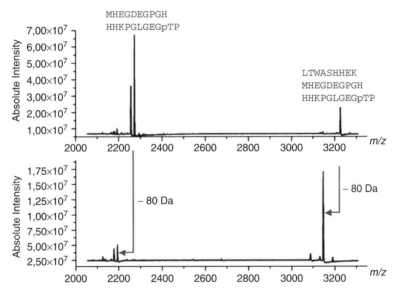

FIGURE 14 Chelate complex formation of a phosphate group with immobilized metal ions. The chromatography can be performed with iminodiacetic acid or the tridentate nitrilotriacetic acid. (Modified from a schematic generated by M. Zeller during his Ph.D. work, Münster, Germany, 2004.) (*See insert for color representation of figure.*)

FIGURE 15 MALDI-TOF spectrum of IMAC-enriched MRP14/MRP14* (myeloid-related protein) digest before and after dephosphorylation. Shown are two phosphopeptides carrying phosphothreonine, with one being one tryptic peptide longer than the other (one missed cleavage). (Modified from spectra generated by M. Zeller during his Ph.D. work, Münster, Germany, 2004.)

Mass spectrometry allows us to detect marker ions that are characteristic for specific groups. In the case of phosphorylation the instrument can be set to look for the diagnostic ions $H_2PO_4^-$ (*m/z* 97), PO_3^- (*m/z* 79), and PO_2^- (*m/z* 63) in negative-ion mode in a data-dependent manner (Figure 16). Also, the immonium ions may be of value for that type of experiment. In particular, analysis-resistant cases of

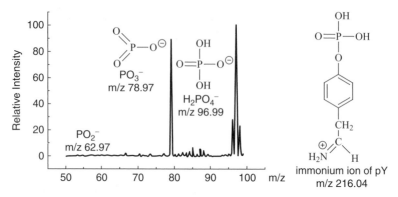

FIGURE 16 Diagnostic ions for phosphopeptides in negative-ion mode. (Modified from spectra and schematics generated by M. Zeller during his Ph.D. work, Münster, Germany, 2004.)

derivatization of phosphopeptides may be considered (Figure 17). β-Elimination and Michael addition creates more stable derivatives, which in addition form low-energy collision-induced dissociation marker ions.

The ultimate goal of many modification analyses, including phosphorylation, is the generation of a tandem MS spectrum of the modified peptide in order to assign the site. As for most protein modifications, MS can deliver data in different

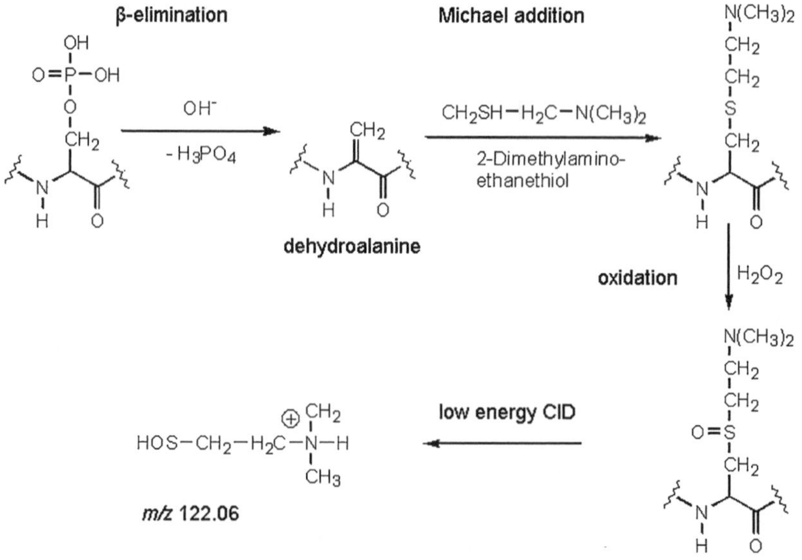

FIGURE 17 β-elimination and Michael addition of pS. The derivatization is also applicable to pT, but not to pY. (Modified from spectra and schematics generated by M. Zeller during his Ph.D. work, Münster, Germany, 2004.)

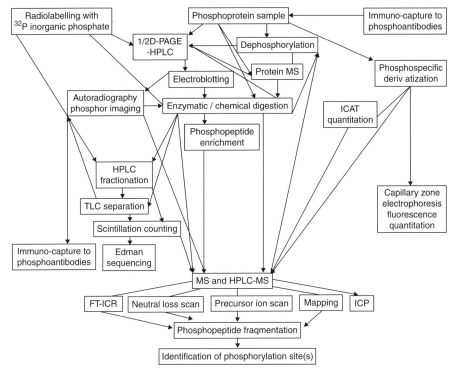

FIGURE 18   Schematic of possible experimental proceedings to identify phosphorylation sites. For an overview, see Zeller and Konig (2004).

experimental settings (Figure 18). However, analysis is seldom straightforward, and phosphopeptides, particularly, show a tendency to resist detection, so that several approaches may be necessary.

It should be mentioned that gel-based methods used to track modifications have their merit in modern proteomics. An example is the differential display of phospho-proteins using phosphogroup-specific staining (e.g., ProQ Diamond). This approach may serve to follow the stimulation of cells or the effect of phosphatases.

DISULFIDE BRIDGES AND CROSS-LINKS

Some protein modifications affect two peptide chains at the same time. These are disulfide bridges or biologically or chemically induced cross-links. A common linking reaction in protein biochemistry uses the EDC (Figure 19), which cross-links carboxyl groups to primary amines or hydrazides. Analysis of intra- and intermolecular cross-links requires knowledge of the linking reaction and dedicated approaches. Since the proteolytic linked peptides can become comparatively large, they may not be

1-Ethyl-3-(3-Dimethylaminopropyl)-carbodiimide Hydrochloride
(EDC)

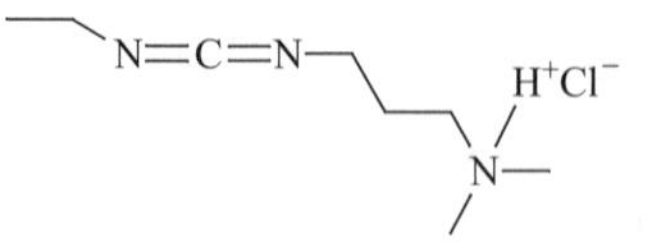

couples carboxyls to
primary amines or hydrazides

EDC Coupling Reaction

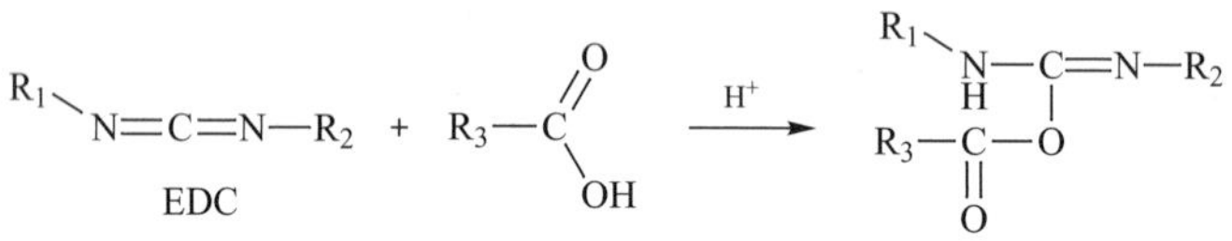

EDC reacts with carboxylic acid group and activates the carboxyl group, allowing
it to be coupled to the amino group in the reaction mixture.

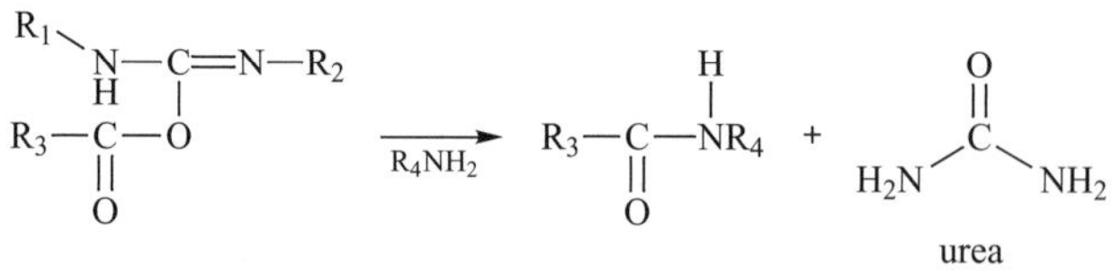

EDC is released as a soluble urea derivative after displacement by the
nucleophile, R_4NH_2

FIGURE 19   Chemical cross-linking using the homobifunctional agent EDC. (Modified from
a Pierce catalog, 1994.)

easily accessible for collision-induced dissociation. The use of different enzymes
with complementary specificity may assist in such cases.

CHEMICAL- AND HANDLING-INDUCED MODIFICATIONS

Deamidation of asparagines residues (Figure 20) is one of the most common modifications in therapeutic proteins produced using recombinant DNA technology. Through
this process, asparagine residues are converted to isoaspartate and aspartate. Methionine oxidation is regularly observed due to handling processes and can hardly be
avoided.

Reduction and alkylation are procedures performed regularly in protein separation.
They are an established part of the two-dimensional gel electrophoresis protocol.
Iodoacetic acid or iodoacetamide is used to alkylate reactive cysteine residues to
prevent re-formation of disulfide bonds after reduction with dithiothreitol (Figure 21).
Urea is often used in the process to facilitate the denaturation and solubilization of

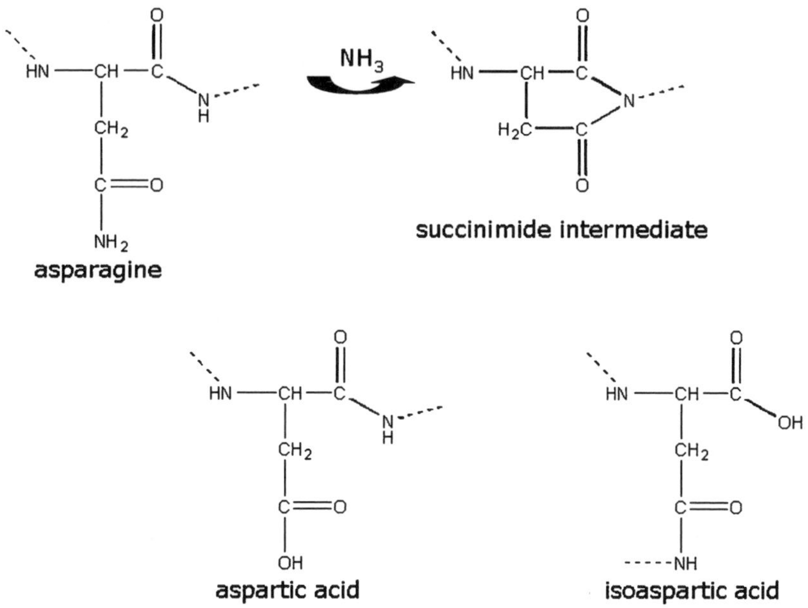

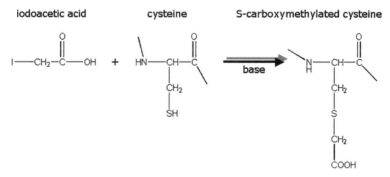

FIGURE 20   Deamidation of asparagines residues.

proteins. As urea in solution is in equilibrium with ammonium cyanate, carbamylation of proteins may occur (Figure 22).

The generation of methyl esters is a reaction that allows us to obtain structural information: for instance, on the number of acidic residues in unknown sequences. Each methyl group adds 14 Da and reacts with the C-terminus. Another helpful analytical reaction is the acetylation of amino groups of N-termini and the side

FIGURE 21   S-Carboxymethylation of the amino acid residue cysteine with the alkylating agent iodoacetic acid.

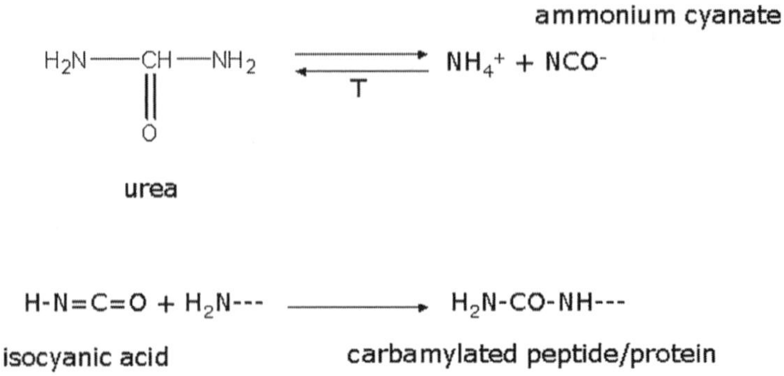

ammonium cyanate

$$H_2N\!-\!\!-\!\!-\!CH\!-\!\!-NH_2 \;\rightleftharpoons\; NH_4^+ + NCO^-$$

$$\overset{||}{O}$$

$$T$$

urea

$$H\!-\!N\!=\!C\!=\!O + H_2N\!-\!-\!- \;\longrightarrow\; H_2N\!-\!CO\!-\!NH\!-\!-\!-$$

isocyanic acid   carbamylated peptide/protein

FIGURE 22   Carbamylation of proteins due to contact with urea. If the temperature is maintained at higher levels for some time ammonium cyanate/isocyanic acid is formed, which reacts with the N-terminus, lysine, arginine, and cysteine.

chain of lysine. It allows us to count the number of lysine residues or to distinguish glutamine and lysine, which have nearly isobaric mass. Acetylation introduces a mass shift of 42.0106 Da.

DATABASES

Databases such as RESID have been assembled in the public domain to collect the state of knowledge. For the special needs of mass spectrometry, Unimod is being generated to meet the following criteria: (1) to be comprehensive and contain biological and artificial modifications as well as artifacts due to sample handling, (2) to provide both monoisotopic and average molecular mass, (3) to deliver sequence or site information in addition to constraints and links to further data, and (4) to contain MS-relevant information such as neutral losses or immonium ions.

CONCLUSION

Modification analysis, in particular with respect to protein function, requires experience and patience. New modifications are still being discovered. For an overview on the current knowledge, see Table 3.

Acknowledgments

For knowledge on chemically induced modifications, the Web site ionsource.com is gratefully acknowledged.

TABLE 3 Protein Modifications

Modification	Abbreviation	Kingdom	Residues/Rule	Mass Difference Monoisotopic/Average	UniProtKB/Swissprot Annotation
			Cysteines		
Cysteine sulfenic acid (-SOH)	CSEA		C	15.9949146/ 15.9994	
Cysteine sulfur acid (-SO$_2$H)	CSIA	Metazoa, bacteria, plants?	C	31.9898292/ 31.9988	
Cysteine persulfide	CYSP		C	31.9721/32.06	
Glutathionylation	GLUT	Metazoa, bacteria	C	305.0680814/ 305.3056	Glutathione
S-12-Hydroxyfarnesyl cysteine	FARO		C	220.1827/220.35	
FMN conjugation (Cys)	FMNC		C	456.1046/456.34	S-4a-FMN cysteine, S-6-FMN cysteine
Pyruvic acid	PYRUC		C	−33.0037/−33.10	
S-Nitrosylation	NTRY		C	28.99017/28.99816	S-Nintrosocysteine
3-Oxoalanine (Cys)	OXOAC		C	−17.9928/−18.08	
S-Palmitoleyl cysteine	PALE	Yeast	C	236.2140/236.39	
S-Diacylglycerol cysteine	DIAC	Prokaryotes, phages	C and N-termini after cleavage of signal peptide, frequently associated with N-palmitoyl cysteine (palmitoylation)	576.5117/576.94	S-Diacylglycerol cysteine, representative structure: palmitate- and oleate-substituted glycerol
S-Archaeol	ARCH	Archaea	C on N-terminus after cleavage of signal peptide, frequently associated with N-acetyl cysteine (acetylation)	634.6628/635.14	S-Archaeol cysteine

(Continued)

TABLE 3 Protein Modifications (*Continued*)

Modification	Abbreviation	Kingdom	Residues/Rule	Mass Difference Monoisotopic/Average	UniProtKB/Swissprot Annotation
S-Farnesyl cysteine	FARN	Eukaryotes, viruses	C, three residues away from C-terminus, PDOC00266, frequently associated with cleavage of a C-terminal propeptide (CAAX processing) and subsequent methylation of new C-terminus	204.1878/204.3556	
Geranyl geranylation	GERA	Eukaryotes, viruses	C, three residues away from C-terminus, PDOC00266, frequently associated with cleavage of a C-terminal propeptide (CAAX processing) and subsequent methylation of new C-terminus	272.2504/272.4741	S-Geranylgeranyl cysteine
			Acids/Amides		
β-Methylthiolation	BMTH	*E. coli*	D	45.9877118/46.08688	3-Methylthioaspartic acid
cis-14-Hydroxy-10,13-*d* ioxo-7-heptadecenoic acid aspartate ester	CHDH	Plant	D	294.1831/294.39	
Deamidation	DEAM	All	Q anywhere, N anywhere, but needs to be followed by a G	0.9840/0.9847	Deamidated asparagine/glutamine
Deamidation followed by methylation	DEAME	All	Q	14.9997/15.01	Glutamate methyl ester (Gln)
Pyrrolidone carboxylic acid	PYRR	All	N-Terminal Q	−17.0266/−17.0306	

Name	Abbrev	Organisms	Amino acid / consensus	Mass	Representative structure
Pyrrolidone carboxylic acid (Glu)	PYRE		N-Terminal E	−18.0106/18.02	
γ-Carboxyglutamic acid	GGLU	Eukaryotes	E	43.98983/44.0098	4-Carboxyglutamate
ω-Hydroxyceramide glutamate ester	OHC	Eukaryotes	E	760.7308/761.3	Representative structure: triacontanoate-substitute *d*-icosasphingosine

Lysines

Name	Abbrev	Organisms	Amino acid / consensus	Mass
Allysine	ALLYS		K	−1.0316/−1.03
N^6-Poly (methylamino-propyl))lysine	POLYM		K	426.4410/426.73
Pyrrolysine	PYRK		K	109.0528/109.13
N^6,N^6,N^6-Trimethyl-5-hydroxylsine	TRIMETK		K	59.0497/59.09
Hypusine	HYPU		K	87.0684/87.12
N^6-1-Carboxyethyl lysine	CETH		Side chain of K	72.0211/72.06
Biotin	BIOT	All	K, PROSITE PDOC00167 consensus pattern: [GN]-[DEQTR]-x-[LIVMFY]-x(2)-[LIVM]-x-[AIV]-M-K-[LMA]-x(3)-[LIVM]-x-[SAV]	226.0776/226.2934
Lipoyl	LIPY	All	K, PROSITE PDOC00168 consensus pattern: [GN]-x(2)-[LIVF]-x(5)-[LIVFC]-x(2)-[LIVFA]-x(3)-K-[STAIV]-[STAVQDN]-x(2)-[LIVMFS]-x(5)-[GCN]-x-[LIVMFY]	188.033/188.3027

(Continued)

TABLE 3 Protein Modifications (*Continued*)

Modification	Abbreviation	Kingdom	Residues/Rule	Mass Difference Monoisotopic/Average	UniProtKB/Swissprot Annotation
			Alkylation		
Acetylation	ACET	All	K anywhere, N-Terminal not N, K, R, H, F, W, Y	42.0106/42.0373	N-Acetylalanine/-aspartate/ cysteine/glutamate/glycine/ methionine/proline/serine/ threonine/tyrosine/valine, N^2-acetylarginine, N^6-acetyllysine
Methylation	METH	All	C, K, R, H, D, E, N, Q anywhere, but not N-terminus	14.0157/14.0269	5-Methylarginine, 2-methylglutamine, cysteine/glutamate (Glu)/leucine/lysine methyl ester, Methylhistidine, N-methylalanine/isoleucine/ leucine methionine/phenylalanine/ tyrosine, N^4-methylasparagine, N^5-methylarginine/glutamine, N^6-methyllysine, ω-N-methylarginine, Pros-methylhistidine, S-methylcysteine, tele-methylhistidine
Dimethylation	DIMETH TRIMETH			28.0314/28.0538	Asymmetric dimethylarginine N^4, N^4-Dimethylasparagine, N^6, N^6-dimethyllysine, ω-N-methylated

Name	Abbrev	Taxonomy	Site / Consensus	Mass (mono/avg)	Residue
		Eukaryotes	N-Terminal P or A		
		Prokaryotes	N-Terminal A		
			F, Y, I, L, M N-terminal after cleavage of signal peptide; PS00409 consensus pattern, [KRHEQSTAG]-G-[FYLIVM]-[ST]-[LT]-[LIVP]-E-[LIVMFWSTAG]		
Dimethylation of proline	DIMETP	All	P	29.0391/29.06	*N,N*-Dimethylproline
Other					
Pyridoxal phosphate	PLP	All	K, no D, P in pos. −1; no E, P in pos. +1	229.014/229.129	
Hydroxylation	HYDR		P, K, D, N	15.9949/15.9994	3′,4′,-Dihydroxyphenylalanine, 3-hydroxyasparagine/aspartate/proline/ tryptophan, 4-hydroxyarginine/ proline, 5-hydroxylysine, hydroxyproline
Methionine sulfoxide			M	15.9949/15.9994/	
Methionine sulfone	MSONE		M	32.00/31.9898	
FMN conjugation	FMNH		H	454.0889/454.33	Tele-8α-FMN histidine

175

TABLE 3 Protein Modifications (*Continued*)

Modification	Abbreviation	Kingdom	Residues/Rule	Mass Difference Monoisotopic/Average	UniProtKB/Swissprot Annotation
Diphthamide	DIPH		H	143.1184/143.21	
FAD	FAD	All	C, H, Y	783.1415/783.542	O-8α-FAD tyrosine, Pros-8α-FAD histidine, S-8α-FAD cysteine, Tele-8α-FAD histidine
Palmitoylation	PALM	Eukaryotes	S, T, K	238.2297/238.4136	N-Palmitoyl cysteine, N^6-palmitoyl lysine, O-palmitoyl serine/ threonine, S-palmitoyl cysteine
			C, N-terminus after cleavage of signal peptide, associated with cholesterol glycine ester, C		
			C-terminal part of mature peptide		
		Prokaryotes	K		
			C, N-terminus after cleavage of signal peptide, frequently associated with S-diacylcholesterol cysteine		
Dihydroxylation	DIHYDR		R, P, K	31.9898/32.00	3,4-Dihydroxyarginine/ proline, 4,5-dihyroxylysine
Citrullination	CITR	Mammalia	R	0.9840276/0.98476	Citrulline
3-Phenyllactic acid	FLAC		F	0.9840/0.98	
3′,4′-Dihydroxy-phenylalahine	DOPA		F	15.9949/16	
Bromination	BROM	All	Side chain of H, F, W	77.9105/78.9	Bromohistidine, 6′-bromotryptophan

Name	Abbr.	Organism	Residue/Location	Mass (mono/avg)	Representative structure
1-Thioglycine	THIOG		G	15.9772/16.06	
Phosphatidyl ethanolaminie amidated glycine	PE	Eukaryotes	G, C-terminus after cleavage of C-terminal peptide	699.5203/699.98	Representative structure: palmitate- and oleate-substituted glycerol bound to phosphoethanolamine
Cholesterol	CHOL	Eukaryotes	G, C-terminus of N-product after cleavage of C-product	368.3443/368.64	Cholesterol glycine ester
Myristoylation	MYRI	Eukaryotes viruses	G, N-terminus after removal of initiator methionine, PDOC00008	210.1984/210.3598	N-Myristoyl glycine N^6-myristoyl lysine, S-myristoyl cysteine
ADP ribosylation	ADP		E, R, C, S, N	541.0610/541.30	ADP-ribosylarginine/asparagines/cysteine/serine, polyADP-ribosyl glutamic acid (polymer of ribosylglutamic acid)
Formylation	FORM	All	All residues on N-terminus	27.9949/28.0104	N-Formylmethionine/glycine, N^6-formyllysine
Amidation	AMID	Eukaryotes	All residues on C-terminus, needs a G after C-terminus on precursor	−0.9840/−0.9847	Aspartic/glutamic acid 1-amide
Hydroxyl Groups					
2′,4′,5′-topaquinone	TOPA		Y	29.9742/29.98	
Thyroxine	THRX		Y (side chain)	595.6123/595.68	
Triiodothyronine	THRN		Y (side chain)	469.7158/469.78	

(Continued)

TABLE 3 Protein Modifications (*Continued*)

Modification	Abbreviation	Kingdom	Residues/Rule	Mass Difference Monoisotopic/Average	UniProtKB/Swissprot Annotation
(Z)-2,3-Didehydro-tyrosine	DHY		Y	−2.0156/−2.02	
2-Oxobutanoic acid	OXOB		T	−17.0265/−17.03	
3-Oxoalanine (Ser)	OXOAS		S	−2.0156/−2.02	
Phosphopantetheine	PPAN		S, PROSITE PDOC00012: [DEQGSTALMKRH]-[LIVMFYSTAC]-[GNQ]-[LIVMFYAG]-[DNEKHS]-S-[LIVMST]-{PCFY}-[STAGCPQLIVMF]-[LIVMATN]-[DENQGTAKRHLM]-[LIVMWSTA]-[LIVGSTACR]-x(2)-[LIVMFA]	339.078/339.3234	
2,3-Didehydroalanine	DHAS		S	−18.0106/−18.02	
2,3-Didehydrobutyrine	DHB		T	−18.0106/−18.02	
n-Decanoate	DECA		S, T	241.1678/241.33	O-Decanoyl serine/threonine

Name	Code	Organism	Residue	Mass	Description
FMN conjugation	FMN		S, T	438.0940/438.33	FMN Phosphoryl serine/threonine
n-Octanoate	OCTA		S, T	126.1044/126.1986	O-Octanoyl serine/threonine
Sulfation	SULF	Eukaryotes	Y, S, T	79.9568/80.0642	Sulfoserine/threonine/tyrosine
Phosphorylation	PHOS		S, T, Y, H, D, C	79.9663/79.9799	4-Aspartylphoshate, Phosphoarginine/cysteine/histidine/serine/threonine/tyrosine, Pros-phosphohistidine Tele-phosphohistidine
Glycosylation					
N-Acetylhexosamine O-GlcNAc	GLCN		S, T, N	203.0794/203.1950	
Glucosylation (glycation; hexose)	GLUC	All	All residues N-terminal; N, T, K anywhere	162.0528/162.1424	
C-Mannosylation	CMAN	Eukaryotes	W	162.052823/162.1424	
Pentose				132.04226/132.1161	
Deoxyhexose				146.0578/146.1430	
Sialic acid				291.0954/291.2579	

Source: http://www.expasy.org/tools/findmod/.

179

SELF STUDY QUESTIONS

Search the literature to answer the following questions.

1. If you had the task of identifying and verifying phosphorylation sites in a protein, how would you proceed?
2. You are supposed to characterize a glycoprotein. What is your strategy? What are the first steps?
3. How would you approach an analysis of disulfide bridges? What difficulties do you expect when using mass spectrometry?

RECOMMENDED READING

Anderson N.L., Anderson N.G. The human plasma proteome: history, character, and diagnostic prospects. *Mol. Cell. Proteom.* 1(2002) 845–861.

Creasy D.M., Cotrell J.S. Unimod: protein modifications for mass spectrometry. *Proteomics* 4(2004) 1534–1536.

Farriol-Mathis N., Garavelli J.S., Boeckmann B., *et al.* Annotation of post-translational modifications in the SwissProt knowledge base. *Proteomics* 4(2004) 1537–1550.

Garavelli J.S. The RESID database of protein modifications as a resource and annotation tool. *Proteomics* 4(2004) 1527–1533.

Hoffmann S., Jonas E., König S., Preusser-Kunze A., Willbold W. Nef protein of human immunodeficiency virus type 1 binds its own myristoylated N-terminus. *Biol. Chem.* 388(2007) 181–183.

McDonald N.Q., Lapatto R., Murray-Rust J., Gunning J., Wlodawer A., Blundell T. New-protein fold revealed by a 2.3-A resolution crystal-structure of nerve growth-factor. *Nature* 354(1991) 411–414.

Robyt J.F. *Essentials of Carbohydrate Chemistry.* Springer-Verlag, New York, 1998.

Stryer L. *Biochemistry* 4th ed. W.H. Freeman, New York, 1996.

Teshima G. http://www.ionsource.com.

Walsh C.T. *Posttranslational Modifications of Proteins.* Roberts and Co. Publishers, Greenwood Village, CO, 2007, Chap. 6.

Zeller M., Vogl T., Roth J., König S. Phosphorylation site analysis of MRP14 purified from granulocytes. *Biomacromol. Mass Spectrom.* 1 (2) (2007) 127–133.

Zeller M., König S. Impact of chromatography and mass spectrometry on the detection of protein phosphorylation sites. Invited review. *Anal. Bioanal. Chem.* 378(2004) 898–909; Erratum, *ABC* 379(2004) 318.

9

MICROFLUIDICS

AGNIESZKA KRAJ AND JERZY SILBERRING

Analytical systems integrated and miniaturized on silica-, glass-, or polymer-based substrates are becoming more common in the proteomic area. Complete ready-to-work analytical systems called lab-on-a-chip or micro-total analysis systems (µTAS) can be smaller than a mobile phone, although the entire instrument of which it is a part is much larger. The rather enthusiastic perspectives of nanoscale analysis of complex samples have been slowed down in recent years, as microfabrication of devices characterized by high throughput is still under early development. The original idea behind miniaturized, integrated analyzers was to produce cheap, easy available, disposable detection systems for a defined set of compounds for diagnostic purposes (something more efficient than a pregnancy test). Challenges of the recent decade extended those perspectives for application of microdevices in high-throughput proteomics and metabolomics. At the time of this writing we do not see any reliable commercial instrument that fulfills large-scale proteomic demands. There are, however, less efficient systems capable of analyzing single molecules at high sensitivity and in a very short time. From a technical point of view, this is probably the way that such development of complex devices should follow, as in any other case.

A basic question we should ask ourselves is: What feature of a given microdevice cannot be achieved using a capillary column or standard capillary zone electrophoresis (CZE) instrument? Then we should follow an excellent review by Lion et al. (2003) in which problems that should be considered at the early phase of design and application of miniaturized devices (not necessarily in proteomics) are discussed. Besides the high-throughput problems and coupling to a mass spectrometer, there is another key parameter: sensitivity. According to calculations given by Lion et al., injection of a picoliter sample requires detection of 10^{-21} mol. Moreover, it seems useless to design a microspray emitter smaller than 10^{-15} L because analytes will never be detected by MS. Even with preconcentration by a factor of 10^3 preceding analysis, there are still attomoles to deal with. There are publications showing spectacular achievements in this field, but transfer of knowledge from such leading institutes to routine laboratory work will take time, despite the availability of really skillful scientists.

Microfluidics devices are able to handle tiny amounts of liquid (of picoliter range) transported in channels nanometers in inside diameter. The technique is taken from photolitography and electronics; thus, all structures can be minimized. The main drawbacks are:

- Sample evaporation
- Sample use (small volumes and sample reservoirs)
- Flow maintenance
- Very clean reagents and sample to avoid clogging
- Connection to detector (e.g., mass spectrometer)
- Difficulties in mimicking column separation
- Low throughput capabilities

Those problems are typical for the "development age" of such systems and are being improved continuously. To show the positive side of this technology, we also list the most important advantages:

- Reduction in reagent consumption and waste
- Reduction in cost per analysis
- Short analysis time
- Increased resolution
- Minimized and simplified operation

MICROFLUIDIC APPLICATIONS IN PROTEOMICS

Protein Arrays

Protein arrays are not real microfluidic devices but miniaturized assay systems. Arrays can also consist of peptides or oligonucleotide probes. They serve, for example, for rapid screening for protein–protein interactions or for detection of gene mutations or gene activity patterns. The arrays can simultaneously detect molecules of interest in complex mixtures, and this is an approach that can be classified as a part of proteomics strategy. The common feature of the arrays is their flat surface, making them an ideal target for direct analysis in a MALDI-TOF instrument.

An interesting attempt to utilize this idea was published with chips used in the surface plasmon resonance method (Biacore). Protein complexes are first analyzed in the SPR instrument and then coated with MALDI matrix and transferred to the mass spectrometer for final identification of the interacting molecules. Several manufacturers offer coated plates (surface chemistry) that can be used for preparation of their own arrays.

INTEGRATED MULTIPLE NANOSPRAY CHIP

The idea behind on-chip ion production has resulted in many promising solutions. A recent development has been described by Jack Henion and co-workers, utilizing a silicon-based ESI chip device. The chip consists of 100 electrospray emitters (nozzles) and can analyze 100 samples in sequence automatically. It is also possible to link liquid chromatography to the chip. Based on this discovery, the company Advion BioSciences was founded in 1993 to manufacture a complete robotic system called NanoMate. The key features of such devices is that they can be manufactured with high reproducibility, are disposable, and can, in many cases, replace troublesome preparation of capillary emitters, which need to be coated with conductive material.

ESI on a Chip

Interfacing chip technology with an electrospray ion source has been investigated almost simultaneously with development of the capillary electrophoresis (CE) chip. There are several solutions, and Figure 1 shows an example of the commercially available HPLC chip, which can be linked to ESI-MS. The chip is also integrated with a sprayer. The disadvantage of such solutions in most cases is the incompatibility of the chips with instruments from other manufacturers. The advantage is a plug-and-play principle, minimizing operation time and simplifying the entire analysis.

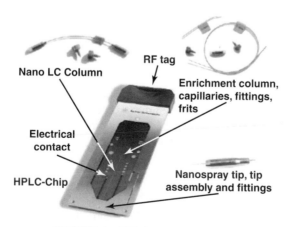

FIGURE 1   HPLC chip from Agilent.

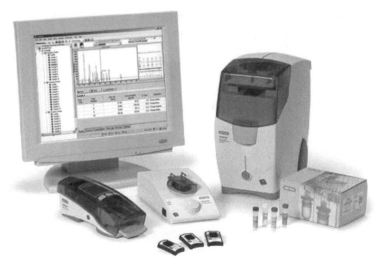

FIGURE 2   Electrophoresis apparatus with chip technology.

Sample cleanup on microfluidic devices by wall derivatization (using a compound with the functionality of a chromatographic stationary phase) or microparticles (difficult in terms of filling of nanochannels), based on monolithic material and polymeric layers.

Electrophoresis on a Chip

Capillary electrophoresis was probably the first technique in which chip technology was used. This was a natural development, as CE utilizes separation in a liquid without a stationary phase. Figure 2 shows a separation system with the trade name Experion, produced by BioRad. The major feature of this system is that the company has also developed stationary phases that can be loaded on demand. This approach attempts to convert tiny channels into a powerful separation device. Alternatively, monolithic materials can be used in the preparation of a chromatographic chip with various stationary phases.

It is clear that miniaturization in proteomics and in analytical sciences is one of the most important and promising goals for the next decade. On the other hand, we must conclude (with a bit of irony) that the only effective high-throughput microfluidic system today is a well-known MALDI spotter.

Acknowledgments

This work was supported by International Centre for Genetic Engineering and Biotechnology (ICGEB) grant CRP/POL05-02.

RECOMMENDED READING

Lion N., et al. Microfluidic systems in proteomics. *Electrophoresis* 24(2003) 3533–3562.

Nedelkov D., Tubbs K.A., Nelson R.W. Surface plasmon resonance-enabled mass spectrometry arrays. *Electrophoresis* 27(2006) 3671–3675.

Zhang S., Van Pelt C.K. Chip-based nanoelectrospray mass spectrometry for protein characterization. *Expert Rev. Proteom.* 1(2004) 449–468.

10

PROTEOMICS

10.1

QUANTITATIVE PROTEOMICS

Tomasz Dylag, Anna Drabik, and Jerzy Silberring

To fully understand the roles of proteins in the organism, it is necessary to consider dynamic changes in protein concentrations induced by various environmental factors, pathophysiologal processes, or medications. Classical techniques for quantitation of proteins in multicomponent samples, such as immunoassays [radioimmunoassay (RIA), enzyme-linked immunosorbent assay (ELISA)], may be used to study only one protein at a time. Moreover, they require production of specific antibodies that will still detect various forms of a given protein, provided that they have the same epitope. Therefore, the final result of antibody-based quantitation refers to the sum of all recognized components (referred to as *immunoreactive-like material*). The situation is even worse during analysis of shorter peptides, where cross-reactions, also with nonrelated sequences, are often detected and contribute to the false-positive results.

To determine the influence of various stimuli on protein levels in the proteomic approach, the most important task is detection of differences between two (or several) biological samples, one usually being the control or reference. This is the first test that allows us to judge whether a given protein should be investigated further as a potential marker (another issue is the level of such difference between protein levels sufficient to consider an up- or down-regulated protein molecule as a biomarker).

> Quantitative proteomics aims at simultaneous quantitation of level differences between many proteins in different samples, not at measurement of their absolute concentrations.

In comparative proteomics, each protein and its role in pathophysiology must be evaluated using various methods, and "normal" and "pathological" values should be estimated with good statistics and on large populations of controls and patients. Moreover, two-dimensional gel electrophoresis or two-dimentional LC-MS must be accompanied at least by Western blotting, which adds value to data collected during preseparation of complex samples and minimizes the risk for false-positive results from the beginning of the analytical process.

Potential errors in quantitative analysis may arise from differences in sample preparation (isolation/extraction procedures, concentration, or fractionation), and procedures of sample analysis (e.g., by chromatography or mass spectrometry). Such errors can be minimized by limiting the number of steps in the protocol and by simultaneous analysis of all samples being compared. In an ideal situation, components from different samples are labeled with distinct tags and combined prior to analysis (such methods are described in this chapter), which could alleviate the problems noted above and also eliminate the risk of possible loss of components in one of the samples.

ELECTROPHORETIC TECHNIQUES

Two-dimensional electrophoresis in polyacrylamide gel (PAGE) is one of the main methods of separation in proteomics. Once separated, proteins are visualized with Coomassie Brilliant Blue, colloidal silver, or a fluorescent marker. Differences between protein levels are quantified by measurement of spot intensities (densitometry). The accuracy of such analysis depends on the dye used for staining, the evaluation software, the spot shape, and the background level, among others. Quantitative proteomics often utilizes fluorescent dyes, as they provide a linear response, which means that signal intensity corresponds accurately to the number of protein copies in a given spot. The features of the popular fluorescence dye SYPRO Ruby are as follows:

- Ease of use
- High sensitivity (1 ng of protein)
- Lack of staining the gel itself (low background)
- Compatibility with MS techniques
- Broad range of linearity (three orders of magnitude)

Therefore, this dye can be used for the determination of small changes over a broad concentration range.

Laboratory practice indicates that it is often difficult to obtain reproducible separations by standard two-dimentional PAGE; therefore, two-dimensional gel electrophoresis was adapted for the needs of quantitative proteomics as Differential gel electrophoresis (DIGE). In this technique, several fluorescent dyes are used to differentially label proteins from distinct samples. The two samples are then combined and analyzed on one gel. Commonly used fluorescent dyes are Cy3 and Cy5 (Figure 1). Cy3 is excited at 550 nm, Cy5 at 649 nm; they emit light at 570 and 670 nm, respectively. Figure 2 presents the principle of DIGE. The samples are processed together under identical conditions; therefore, superposition of two electropherograms recorded at different wavelengths and their quantitative comparison is relatively easy.

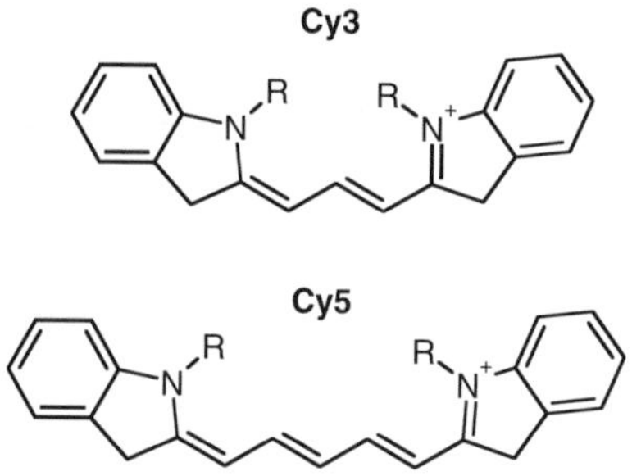

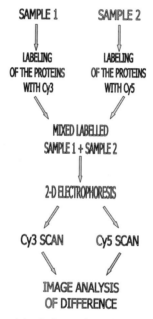

FIGURE 1   Structures of Cy3 and Cy5. R denotes aliphatic chains that include moieties such as maleimide or *N*-hydroxysuccinimide residues that react with proteins.

FIGURE 2   Differential gel electrophoresis in quantitative proteomics.

MASS SPECTROMETRIC TECHNIQUES

For quantitative analysis, mass spectrometry–based techniques utilize internal standards, that is, compounds that are added to the sample and have properties very similar to those of the components analyzed. In quantitative proteomics, internal standards most commonly include stable heavy isotopes of certain elements and are usually incorporated into sample components as various tags (see Chapter 7.6). The role of internal standard is often played by control sample components, whose peptides are modified by a "light" label, while the components of the test sample are derivatized with a "heavy" label (or vice versa). Both samples are then pooled and, ultimately, analyzed by mass spectrometry. The resulting spectrum shows pairs of peaks whose

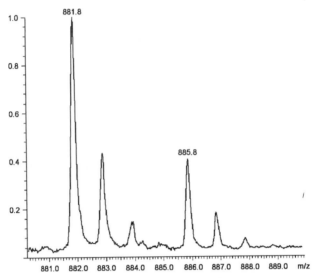

FIGURE 3 Fragment of an ESI mass spectrum of two samples labeled with various isotopic tags with a mass difference of 4 Da.

m/z difference corresponds to the mass difference between the "light" and "heavy" labels. Comparison of the intensities of such peak pairs provides a clue on the relative contents of the respective components in both samples. An example of a peptide map labeled and analyzed by LC-MS is shown in Figure 3.

Labeling with light and heavy isotopes can be performed either in vitro, after isolation of proteins from biological material, or in vivo in living cells, as described further in this chapter. Several labeling procedures prior to mass spectrometry are shown in Figure 4. One of the problems concerning labeling with isotope tags is a huge number of peptides generated during proteolysis, far exceeding several thousand. Therefore, it is advisable to utilize more efficient separation methods, such as two-dimensional LC-MS, instead of standard one-dimensional LC-MS.

In Vivo Labeling

In vivo methods are based on incorporation of light or heavy isotopes into proteins present in living cells. These methods are characterized by high confidence of measurements, as proteins are labeled and samples are combined *before* extraction of proteins from biological material. All subsequent operations are performed on the mixture of samples, which minimizes the risk of error.

Cell cultures are widely used in proteomic research, as they are easy to control. To compare two cell populations, one culture is grown in a standard medium and the other in a medium enriched in stable ^{15}N nitrogen isotopes. They are incorporated into all proteins present in the cells. Now, both cell populations are pooled and lysed. Proteins are extracted, digested, and analyzed by LC-MS. Relative intensities of peak

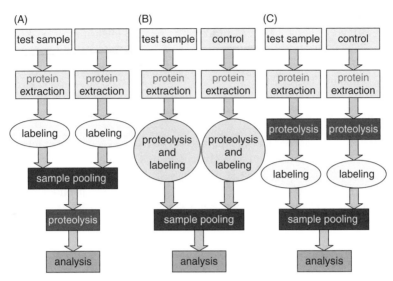

FIGURE 4   General scheme describing procedures that involve isotope tags and mass spectrometry: (A) in vivo labeling; (B) in vitro labeling prior to proteolytic digestion; (C) in vitro labeling of peptide map. (*See insert for color representation of figure.*)

doublets (light/heavy) are proportional to the content of particular peptides, and so also to the content of proteins from which these peptides derive.

An important in vivo approach in cell culture research is stable isotope labeling with amino acids in cell culture (SILAC). Different cells are grown in specific media containing distinct forms of a particular amino acid (e.g., hydrogen-containing and deuterated leucine). The amino acid is selected carefully, as it should be exogenous, so that the cells will take it up from the medium. Following proliferation, the content of a labeled amino acid reaches nearly 100% in a cell population. Then the two differentially labeled populations (i.e., one labeled with standard leucine and the other labeled with deuterated leucine) can be combined and compared.

The major drawback of this and other techniques based on deuterated compounds is a difference in retention times on the chromatographic column, compared to nondeuterated compounds. Deuterated peptides are slightly more hydrophilic and are eluted earlier from the reversed-phase column. This effect increases as more deuterated amino acid residues are present in the peptide sequence. In vivo labeling of proteins present in multicellular organisms becomes difficult and very expensive. Therefore, such methods are not used in animal studies.

In Vitro Labeling

The number of possibilities of incorporating isotope tags in vitro (i.e., after extraction of proteins from biological material) is much higher than with in vivo methods. The simplest techniques are based on esterification of carboxyl groups or acetylation of amino groups. Proteins are proteolytically digested and the resulting peptides are derivatized, pooled, and analyzed further. The relative intensities of peak pairs are then compared. The major advantage of this approach is the possibility of application to all samples, regardless of their origin (e.g., cell homogenates, tissues). More details may be found in Chapter 7.6.

Proteolytic Labeling

Proteolytic labeling is based on incorporation of various oxygen isotopes into peptides or proteins that are subjected to proteolytic digestion. Depending on the sample (test or control), enzymatic digestion is performed in either $H_2{}^{16}O$ or $H_2{}^{18}O$. Oxygen isotopes are incorporated into the carboxyl terminus of the peptides released. A disadvantage of this method is a possible overlap of isotopic patterns. This approach is described in detail in Chapter 7.6.

ICAT

The ICAT method (isotope-coded affinity tag) is based on a chemical reaction of proteins with the reagent, consisting of the following components:

- Iodoacetamide residue, which binds to cysteine thiol group
- Linker containing hydrogen or deuterium atoms
- Biotin residue, which permits isolation of the labeled molecules by affinity chromatography

Samples to be compared are incubated with light or heavy reagents, as shown in Figure 5. ICAT reagent binds to the –SH groups of cysteine residues via the iodoacetamide moiety. After the reaction has been completed, the two samples are combined and digested with trypsin. Cysteine-containing peptides are isolated from the mixture by affinity chromatography by the presence in the label of biotin, which is able to interact with avidin immobilized on a stationary phase. Further steps involve LC-MS analysis. Peptide pairs differ by 8 Th in the spectrum, and the intensities of respective peaks can be compared. The advantages of ICAT are:

- High specificity
- High sensitivity
- Applicability to samples of different origin (cells, tissues, fluids)
- Effective labeling in the presence of guanidine, SDS, or urea

Labeling of fragments containing cysteine residues simplifies separation significantly, as all other proteins are excluded from analysis, and makes the spectra easier to interpret. On the other hand, it does not allow for quantitation of proteins that do not contain Cys residues (ca. 10% of proteins). Another drawback is that the mass of

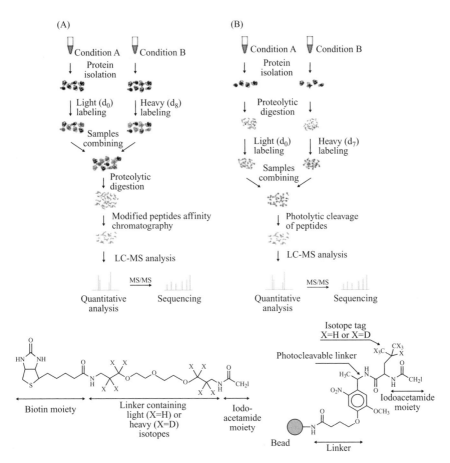

FIGURE 5   Scheme of an ICAT experiment: (A) reaction in liquid; (B) solid-phase ICAT. The structures of the chemicals are shown at the bottom. (*See insert for color representation of figure.*)

the label itself exceeds 400 Da; therefore, mass determination may be less accurate due to decreasing resolution, although this becomes less significant with modern mass spectrometers.

A solid-phase alternative has been developed (solid-phase ICAT) in which proteins are digested by trypsin before derivatization. The ICAT reagent (leucine derivative) is immobilized on the solid support via a UV-sensitive bond (Figure 5B). Peptides are derivatized with corresponding light or heavy labels, pooled, and the components that do not bind to the solid support are washed away. Bound peptides are released from the support by UV irradiation and analyzed further by LC-MS. The solid-phase ICAT procedure is much faster and easier to conduct, as it does not require isolation of labeled peptides by chromatographic methods. All peptides that reach the stage of

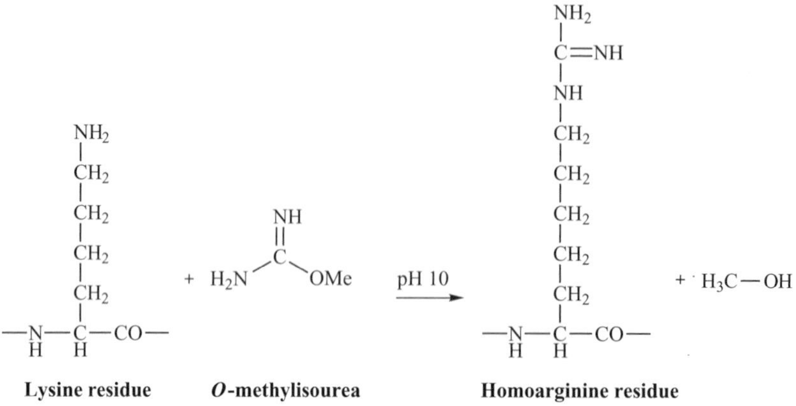

FIGURE 6   Lysine guanidination with O-methylisourea.

LC-MS analysis are modified by the leucine derivative, which is much smaller than the solution-phase ICAT label.

MCAT

The MCAT (mass-coded abundance tag) method is based on guanidination of the ε-amino groups of the lysine residues of tryptic peptides (peptides that are formed during proteolytic cleavage of proteins with trypsin; in principle, lysine residues are located on their C-termini). Guanidination occurs upon treatment of such peptides with O-methylisourea at basic pH. As a result of modification, lysine moieties are converted into homoarginine residues (see Chapter 7.6), leading to a mass shift of 42 Da for each modified sample, as presented in Figure 6.

Only one sample is modified with MCAT; the other sample is analyzed without modification. Pairs of ions (native and modified) differ by 42 Da. Similarly, in MS/MS spectra, fragments containing native and modified peptides are separated by the same value. Ions that belong to the b series and fragments without lysine residues remain unchanged. For this reason, comparison of MS/MS spectra of native and modified peptides, identification of the y-ion series, and sequence readout are relatively easy.

Use of the MCAT method is limited to peptides flanked by lysine residue at the C-terminus. Therefore, incomplete enzymatic cleavage of proteins containing lysine may lead to incorrect results.

iTRAQ

iTRAQ (isobaric tags for relative and absolute quantitation) permits simultaneous analysis of up to four samples. The method is based on labeling of peptides with

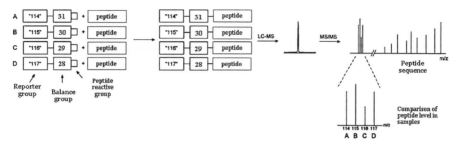

FIGURE 7   Principle of the iTRAQ method.

isobaric tags (i.e., tags that have the same molecular weight), which produce different ions during fragmentation. The iTRAQ tag consists of three components:

1. A moiety reactive toward amino groups (including ε-amino groups of a lysine side chain)
2. A balance group
3. A reporter group

Reporter groups differ in isotopic composition, but due to the balance groups of varying masses, the total weight of the tag is constant and causes a mass shift of 144 Da for each amino group modified. A pooled mixture of all labeled peptides is chromatographed and analyzed in the mass spectrometer. The differences between various tags are visible upon fragmentation, as each fragment spectrum contains one of the characteristic peaks at 114, 115, 116, or 117 *m/z* belonging to the respective reporter groups. The intensity of these ion peaks is a measure of the relative abundance of a particular peptide or protein in a given sample. The principle of iTRAQ is presented in Figure 7. A new version of iTRAQ reagent was introduced recently based on the same principle, which allows simultaneous analysis of up to eight samples.

iTRAQ is a promising tool in quantitative proteomics, although it has several drawbacks. First, fragmentation is necessary to reveal differences in protein content between samples since diagnostic ions are not visible unless MS/MS is performed. Second, ion-trap analyzers have a lower mass exclusion limit during fragmentation, and the diagnostic ions fall below that threshold. Therefore, to benefit from the iTRAQ method, a spectrometer equipped with the quadrupole analyzer may have to be used. This is a good solution but may be time consuming, particularly when nano-LC is used. Alternatively, MS^3 fragmentation and an ion-trap analyzer may be used.

Acknowledgments

This work was supported by International Centre for Genetic Engineering and Biotechnology (ICGEB) grant CRP/POL05-02. Tomasz Dylag received a scholarship from the Foundation for Polish Science.

RECOMMENDED READING

Applied Biosystems. iTRAQ reagents. *Chemistry Reference Guide*. http://www. appliedbiosystems.com, 2004.

Miyagi M., Sekhar Rao K.C. Proteolytic ^{18}O-labeling strategies for quantitative proteomics. *Mass Spectrom. Rev.* 26(2006) 121–136.

Ong S.E., Foster L.J., Mann M. Mass spectrometric-based approaches in quantitative proteomics. *Methods* 29(2003) 124–130.

Schneider L.V., Hall M.P. Stable isotope methods for high-precision proteomics. *Drug Discov. Today* 10(2005) 353–363.

Sechi S., Oda Y. Quantitative proteomics using mass spectrometry. *Curr. Opin. Chem. Biol.* 7(2003) 70–77.

Tao W.A., Aebersold R. Advances in quantitative proteomics via stable isotope tagging and mass spectrometry. *Curr. Opin. Biotechnol.* 14(2003) 110–118.

Whetstone P.A., Butlin N.G., Corneillie T.M., Meares C.F. Element-coded affinity tags for peptides and proteins. *Bioconjugate Chem.* 15(2004) 3–6.

Zhang H., Yan W., Aebersold R. Chemical probes and tandem mass spectrometry: a strategy for the quantitative analysis of proteomes and subproteomes. *Curr. Opin. Chem. Biol.* 8(2004) 66–75.

Zhou H., Boyle R., Aebersold R. Quantitative protein analysis by solid phase isotope tagging mass spectrometry. *Methods Mol. Biol.* 261(2004) 511–518.

10.2

STRUCTURAL PROTEOMICS

Małgorzata Rzychon

Proteins are synthesized as linear polymers that develop biological activity as a result of folding into characteristic three-dimensional structures. Determination of the three-dimensional structures of all existing proteins lies within the scope of research defined as *structural proteomics*. Structural proteomics strongly affects basic research, providing information necessary to understand the relationship between protein sequence, structure, mechanism of action, and function. It has also taken center stage in the biotechnology industry, since structural knowledge of protein domains responsible for biological activity has become a common starting point for drug design.

STRUCTURAL PROTEOMICS PIPELINE

There are a number of key stages in progressing through any structural proteomics project pipeline. The protein targets and suitable cloning strategy are chosen first, followed by selection and optimization of the expression and purification systems. These initial steps, leading to production of the large amounts of pure protein required in structural studies, remain a bottleneck in most structural proteomics projects and are under constant development.

Once protein preparation is available, structural characterization may be started. The most widely used approach is x-ray crystallography, which was employed to solve about 85% of structures deposited in the Protein Data Bank (PDB). Nuclear magnetic resonance spectroscopy (NMR) accounts for about 15% of structural determinations in the PDB, and the contribution of electron microscopy (EM) and atomic force microscopy (AFM) is minimal.

Proteomics: Introduction to Methods and Applications, Edited by Agnieszka Kraj and Jerzy Silberring
Copyright © 2008 John Wiley & Sons, Inc.

The method of x-ray crystallography depends on the production of high-quality crystals, which often represents another bottleneck in structural proteomics pipelines. NMR-specific sample preparation avoids this particular problem but includes isotope labeling at the stage of expression, as well as supplementing purified protein solutions with a number of stabilizing and solubilizing agents in order to obtain samples of reasonable stability at high concentration for the time needed to acquire NMR data.

Regardless of the method chosen, further steps in structural proteomics studies comprise data collection, data processing, and data analysis up to structure determination, refinement, verification, and deposition in the PDB. X-ray crystallography provides structures of very high resolution; however, the data cover only specific conformations of protein molecules that correspond to the organization of crystal analyzed, missing most information on regions that are statically or dynamically disordered. On the other hand, NMR spectroscopy can yield high-resolution data for both ordered and disordered parts of the molecules. The two techniques have advantages and disadvantages and are considered complementary.

The structural proteomics study is assumed complete when the newly determined structure is thoroughly analyzed by comparison with all structures in the PDB, and protein function might be inferred from the data collected (Figure 1).

TARGET SELECTION

Identification of protein targets for structural investigation is a crucial strategic issue. In traditional research laboratories, the choice depends on the researcher's intuition and desire to answer specific questions. However, structural analysis of all existing proteins by this approach would be enormously time consuming. Structural proteomics offers an alternative based on a systematic planning and high-throughput approach, which proved to be useful in initiatives such as the Human Genome Project.

One criterion that contributes to the selection of targets for structure determination is the novelty criterion: the likelihood of identifying a novel protein fold or of solving the structure of a member of a large protein family for which no structural data are available. It is roughly estimated that directing efforts at the experimental determination of 10,000 to 20,000 representative protein structures could result in collecting information on all naturally occurring protein folds. Subsequently, databases constructed on that foundation would enable us to build structural models of the remaining proteins, minimizing the amount of further experimental work required.

The set of proteins to work on is also selected based on their biological or medical significance. For example, there are research projects focused on proteins that are widely distributed throughout different groups of organisms, since such ancient and conserved proteins may represent important biological functions. The others focus on completed genomes and methodically determine or predict structures corresponding to each protein sequence. Yet another approach is to study proteins involved in a single molecular pathway or implicated in a particular medical condition by targeting disease-associated genes. Targets selected under a criterion of biological or medical relevance are then filtered by comparison with sequences already deposited in the PDB, in order to eliminate proteins whose structures are already known.

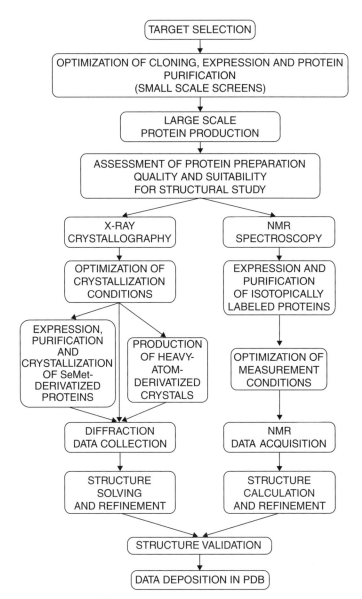

FIGURE 1 Structural proteomics flowchart.

In the target selection procedure for structural studies, technical feasibility is certainly a pivotal factor as well. The important issue to consider is how easily the candidate protein can be produced in the significant amounts required for structural biology investigations. The method of first choice for the production of protein material is the expression in *Escherichia coli*, and a number of combined biochemical data are used to predict which protein or protein fragment will be expressed efficiently and solubly. However, many eukaryotic proteins do not express well in bacteria, and the prioritization of targets according to this criterion means that, for example, those that are predicted to be membrane proteins are less likely to be selected. In general, due to limitations of the available structure determination techniques, the smaller the molecule, the easier and less time consuming are the studies required. This relation, particularly, challenges investigations into eukaryotic organisms, where most genes encode protein molecules composed of numerous globular domains. Since individual domains are often associated with particular biochemical functions, the size problem is overcome by targeting single domains instead of entire molecules. A disadvantage of such an approach might lie in the fact that the structure of the protein as a whole would remain unknown.

SAMPLE PREPARATION

Structural proteomics requires large amounts of high-quality homogeneous protein preparations. As most proteins are, for either technical or ethical reasons, difficult to isolate directly from natural sources, the easiest, quickest, and cheapest approach is represented by protein overexpression in heterologous expression systems. This term stands for directed production of desired protein in large quantity in a foreign organism by using specially designed expression vectors. Briefly, these vectors comprise the DNA fragment encoding the protein of interest, genetic regulatory elements recognized by the expression host organism, and relevant selection markers. Protein overexpression in heterologous expression systems is a topic in itself and will be described only with regard to use in structural proteomics.

In the initial step in the heterologous protein expression pipeline, the DNA containing sequence of interest is isolated directly from the source organism or retrieved from a genomic or cDNA library. A DNA segment of suitable size is then generated either by using restriction endonucleases or by PCR (polymerase chain reaction) amplification, and cloned into the appropriate expression vector.

Selection of the expression vector that is most suitable (i.e., ensuring a high yield of protein production, proper folding, feasibility of scale-up) is best achieved by multiple cloning and expression trials, with the use of a set of different vectors. For easy parallel handling of these numerous experiments, a wide range of cloning methods have been developed that offer flexibility in introducing gene targets into the vectors and their subsequent transfer from one tested vector to another. The three strategies currently most widely used in structural proteomics include traditional ligation-dependent cloning, ligation-independent cloning (LIC), and site-specific recombinatorial cloning (the Gateway system).

The traditional ligation-dependent sticky-end or blunt-end cloning methods utilize ligase, restriction endonucleases, and vectors with multiple cloning sites (MCSs; regions comprising closely spaced sequences recognized by restriction enzymes). The feasibility of subcloning is ensured by providing a set of vectors with nearly identical MCSs. A high-throughput parallelized approach, however, is accomplished more easily by ligation-independent, site-specific recombinatorial cloning.

The LIC technique enables insertion of PCR products containing the target DNA sequence directly into any cloning or expression vector, owing to predesigned primer extensions that match part of the recipient vector sequence. The method utilizes the $(3' \rightarrow 5')$ exonuclease activity of T4 DNA polymerase in the presence of a single deoxyribonucleotide specifically to remove 12 to 15 nucleotides from each $3'$ end of both the PCR product and the linearized recipient vector. The overhangs generated are complementary to one another and therefore anneal strongly enough to neglect in vitro ligation and to transform the host organism with high efficiency. Last but not least, the Gateway technology is grounded in the recombination system of phage λ and involves two steps: incorporation of the PCR product to be cloned into an entry vector by means of recombination, followed by the second recombination event, which transfers the gene of interest from the entry vector into the expression vector, selected from a broad collection of available vectors with the appropriate *att* recombination elements.

The most commonly used heterologous expression system is that of *E. coli*, which involves a variety of vectors with different promoters, multiple fusion tags, and an assortment of protease recognition sequences for tag removal. The greater portion of *E. coli* vectors is based on the T7 polymerase expression system, which is widely adopted because of its high specificity and activity. The protein expressed under the control of the T7 promoter may account for up to 50% of total protein in a bacteria cell in just a few hours after induction. Modern T7-based systems are designed to reduce leaky expression and allow for easy manipulation of the expression level (e.g., by concentration of the added inducer). Moreover, the T7 system is suitable for autoinduction, which makes it particularly useful for fully automated growth of cultures in unique autoinduction media. Bacteria cultures grown in these media do not require monitoring of the optical density or adding any inducer, since they induce production of the target protein spontaneously after reaching a density that is reasonably high.

As the tendency to maximize the efficiency of any procedure is characteristic of structural proteomics, the expression vectors are usually intended to both facilitate the solubility of target proteins and ease their further purification by means of affinity chromatography. Accordingly, the engineered tag sequences are incorporated into the vectors so that the protein expressed is fused to the tag portion, which promotes its folding, enhances stability, or can bind specifically to the relevant chromatographic resin. The most frequently used fusion tags include hexahistidine (6xHis), which enables isolation by immobilized metal chromatography on Ni-NTA beads, a solubility-enhancing thioredoxin (TRX) tag, and both affinity and solubility tags of maltose-binding protein (MBP) and glutathione S-transferase (GST), which can be purified on amylose or glutathione resin, respectively. Following affinity chromatography, the target protein may be separated from its fusion partner by proteolytic cleavage of the peptide bond within the specific recognition site, which was built into the amino acid sequence through an expression vector. Afterward, additional purification step(s) can be carried out if necessary.

The techniques employed by modern molecular biology are prone to automation to allow handling a large number of genes in parallel, with the use of 96-, 384-, or even 1536-well plates and relevant robots or multichannel pipette dispensers. To give an example of a high-throughput sample preparation pipeline: For initial cloning and expression trials, the genes of interest are cloned in parallel into a set of different vectors. Most of the steps implicated in cloning and bacteria cell transformation are conducted in multiwell plates. The transformed colonies are then used to inoculate small-scale cultures, again grown in a multiplate format, followed by plasmid isolation from these cultures, another automated step. Once isolated, the engineered expression plasmids are transformed into the competent cells of a relevant *E. coli* strain and cultured without human intervention in autoinduction media to produce the desired proteins. The cultures are lysed chemically, which is particularly suitable for multiple samples. Next, the affinity-tagged recombinant proteins obtained can all be purified by the same type of chromatography with a liquid-handling robot. The expression levels and solubility are then analyzed by an automated electrophoresis system. The information that results from these small-scale screens is used to identify proteins that can be expressed solubly in bacteria. These protein targets are then expressed on a large scale, purified, and used for structure analysis trials. The targets that fail to be produced efficiently are filtered out and exploited for a second round of screening with an additional set of vectors containing, for example, other solubility-enhancing fusion tags, with different culturing conditions, or even with another expression host. For those proteins that are expressed exclusively in the insoluble fraction of bacteria cell lysate, one of the methods of refolding from inclusion bodies may occasionally be used with success.

Unfortunately, the most commonly used *E. coli* heterologous expression system has numerous limitations, such as the inability to introduce eukaryotic posttranslational modifications and failure to express, or to express solubly, proteins that are large or membrane-associated. Some of these difficulties can be overcome by using a variety of engineered *E. coli* strains that specifically promote the production of troublemakers e.g., proteins with rare codons, membrane proteins, or proteins that may be toxic to the cell. Nevertheless, it is estimated that *E. coli* expression systems will provide structures for about 15 to 20% of naturally occurring small and medium-sized typically nonmembrane proteins. Thus, alternative expression systems are being developed which include yeast, insect, and mammalian cells as well as cell-free translation.

Next most popular after the *E. coli* system is the baculovirus-driven expression in insect cells, which allows for relatively high-level expression of eukaryotic proteins, including the membrane proteins, due to the mechanism of posttranslational processing. Similarly, the yeast systems provide the posttranslational processing machinery, are easy to use, and are suited for large-scale production. On the other hand, the most native environment for the expression of mammalian proteins would be guaranteed by mammalian cell lines; however, such recombinant protein expression is slow, expensive, and nonefficient. Recently developed cell-free expression systems are based on *E. coli* and wheat germ extracts and have proven to be particularly useful for targets that are toxic to living cells. The majority of these expression systems, although very promising, remain at an early stage of technical development with a view to the high throughput required by structural proteomics in either the expression or screening approach.

The final steps in sample preparation for structural proteomics comprise assessment of the quality of the protein material and its suitability for x-ray crystallography or NMR studies. Protein preparation purity, correct folding, monodispersity, and homogeneity are analyzed by mass spectrometry, gel electrophoresis, size-exclusion chromatography, circular dichroism, dynamic light scattering, and one-dimensional NMR. The protein targets are also tested for their stability, oligomeric state, and levels of posttranslational modifications. Quality evaluation allows us to identify proteins that are insufficiently stable, tend to aggregate, or are partially degraded, and either eliminate them from subsequent structural investigations or optimize the purification procedure. Next, the structure determination method is selected.

X-RAY CRYSTALLOGRAPHY

X-ray crystallography is based on deducing the structure of a molecule from the pattern produced by diffraction of x-rays through a lattice of atoms in a crystal. First, extensive screening is set up to identify the most favorable conditions for the growth of target protein crystals of diffraction quality. The variables tested include the type of crystallization buffer used, its pH and ionic strength, the type and concentration of added precipitant, the purity and concentration of a protein to be crystallized, and the crystallization temperature. Typically, the optimal protein concentration is first determined by serial dilutions in a set of specific commercially available solutions and observing the intensity of precipitation. Next, a panel of commercial or homemade screening kits that span a wide range of parameters is set up at two different temperatures (e.g., 4°C and room temperature). Crystals are grown on dedicated crystallization plates, most commonly by a vapor diffusion method, by either a hanging or a sitting drop method (Figure 2). Nowadays, the volumes of buffers used in crystallization trials have been lowered to the nanoliter scale, at the same time reducing the amount of target protein required. Moreover, screen setup and monitoring of the crystallization experiments are both automated using nanoliter-dispensing robots as well as image acquisition and image analysis systems, which allow high-throughput parallel screening of numerous parameters. Once found, the preliminary crystallization conditions are optimized further by setting up more focused screens, followed by scaling-up the procedure to the production of a usable monocrystal of appropriate dimensions and of the highest stability in an x-ray beam (Figure 3).

Optimized diffraction-quality crystals are exposed to x-ray radiation and the diffraction data are collected and processed with specialized software. Currently, the x-ray sources of choice for structural studies of macromolecules are the synchrotron-radiation sources because they deliver radiation thousands of times more intense than conventional x-ray sources, provide diffraction data of much higher resolution, are tunable, and reduce data collection times. To slow down x-ray radiation damage to the crystals analyzed, they are presoaked in a dedicated cryoprotectant solution, frozen in liquid nitrogen, and all measurements then performed at a temperature of about 100 K, maintained by flash-cooling crystals in a nitrogen gas stream.

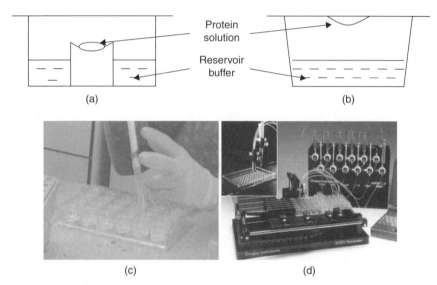

FIGURE 2 Vapor diffusion method of protein crystallization. A droplet containing a protein solution mixed in a 1 : 1 ratio with a solution of precipitant is equilibrated to a reservoir solution that contains precipitant only. (a) Sitting-drop setup; (b) hanging-drop setup; (c) manual preparation of crystallization plates; (d) preparation of crystallization plates with the use of Oryx 6 robot (Douglas Instruments). (Courtesy of S. Odintsov and http://www.douglas.co.uk/oryx.htm.)

The processing of diffraction data collected leads to construction of an electron density map in which maxima correspond to the positions of individual atoms in the unit cell of the target crystal. Unfortunately, structure determination is not a straightforward task, due to a complex mathematical problem known as the *phase problem*, which arises because the measurements provide information on the intensities of the x-ray reflections while phase information is lost.

FIGURE 3 Determination of the structure of staphostatin B (*Staphylococcus aureus* cysteine proteinase inhibitor) by x-ray crystallography: (a) crystal obtained in preliminary crystallization conditions; (b) crystal obtained after optimization of crystallization conditions; (c) diffraction pattern; (d) model of three-dimensional structure. (*See insert for color representation of figure.*)

Numerous strategies are used to solve the phase problem. First, for proteins built from up to several hundred atoms, direct statistical methods are implemented successfully. They are analogous to methods used in small-molecule crystallography and take advantage of limited numbers of atoms in the molecule, which corresponds to the limited number of potential phases. Second, if a suitable model of a target protein or of its large portion can be constructed (e.g., based on the structure of a closely related protein), the missing phase information may be inferred by a computational technique called *molecular replacement*. Finally, although the database of solved macromolecule structures is growing rapidly, yet another approach frequently has to be implemented, which includes isomorphous replacement (IR, heavy atom method) and anomalous dispersion (SAD, single-wavelength anomalous dispersion, and MAD, multiwavelength anomalous dispersion). In the classic method of isomorphous replacement, the heavy atoms (i.e., those with a high atomic number) are incorporated into the target crystal by soaking in a heavy atom solutions such as salts of mercury, arsenic, thallium, osmium, uranium, or bismuth. The resulting crystals are identical to the starting crystals the only exception being that heavy atoms are present. Since x-ray light scattering by heavy atoms is much stronger than light scattering by light atoms, collecting diffraction data from a series of heavy-atom-derivatized crystals provides phase information that is sufficient for deducing the structure of the native crystal. However, soaking in heavy metal solutions is extremely time consuming, due to its low success rate, and at present the most routinely used phasing method is MAD, which uses only one crystal for the entire data collection. In MAD, the anomalously scattering atoms, such as selenium, are incorporated into the target crystal, but instead of soaking of the complete crystal, the recombinant protein is expressed in minimal media in the presence of selenomethionine (SeMet) in order to directly produce SeMet-substituted protein crystals. Since the anomalous scattering power of selenium atoms depends on the radiation wavelength, in a MAD experiment the wavelength of incident x-ray synchrotron radiation is altered to generate a series of diffraction patterns that give information similar to that of isomorphous replacement and allow phasing. Still, the key bottleneck of the technique is the efficient production of SeMet-derivatized crystals, as they do not always crystallize as easily as native crystals.

Subsequent to solving the phase problem and to the construction of an electron density map, an initial model of the target protein structure is built, followed by a computational procedure known as *refinement* that iteratively improves the fit between the experimental data and the model. The refinement is carried out until there is no clear sign of where the model can be improved further. The final structure obtained is then validated by judging a set of statistical criteria and, if satisfactory, the structure coordinates are deposited in the PDB.

Recently, a major effort has been made to develop an automated structure determination using x-ray crystallography. The greatest advances were made in the automation of protein crystallization, allowing fast, high-throughput crystal production. Although significant progress has also been made in the automation of crystal sample handling, characterization, diffraction data collection, and processing, up to model construction, refinement, and validation, there is still much room for further improvement, since the amount of time that has to be dedicated to any individual protein structure remains substantial.

NMR SPECTROSCOPY

Nuclear magnetic resonance (NMR) spectroscopy is an experimental technique that allows us to determine the three-dimensional structures of peptides and proteins in solution (i.e., in conditions comparable to the physiological environment). Since an explanation of the physical basis of NMR is beyond the scope of this chapter, we will just remind the reader that NMR spectra are recorded only for nuclei of atoms that have a nuclear spin and are endowed with an intrinsic magnetic moment, such as ^{1}H, ^{19}F, ^{15}N, ^{13}C, or ^{31}P. Structure determination by NMR is based primarily on correlation between the resonance frequency of each spin and its chemical environment (chemical shifts) and, most gainfully, observation of the nuclear Overhauser effect (NOE), which is due to through-space magnetic dipolar interactions and corresponds to the through-space distance between different spins.

Routinely, the NMR sample preparation starts with isotope labeling of the target proteins with ^{13}C and ^{15}N in order to extend the experimental possibilities from proton-based measurements to more rewarding experiments, including the detection of additional nuclei. The protein targets that can be expressed solubly and efficiently in small-scale expression screens are expressed in the selected host organism grown in minimal media enriched with relevant isotopes, such as uniformly ^{13}C-labeled glucose as the only carbon source and ^{15}N ammonium salts as the only source of nitrogen. In the case of larger molecules, the additional homogeneous or fractional labeling with ^{2}H is often required. Apart from uniform isotope enrichment, structure determination might also be facilitated by selective incorporation of isotopes; however, the straightforward supplementation of growth media with only one labeled amino acid is restricted by undesired metabolic conversion to other amino acids within the expression host organism. This problem may be circumvented by the use of specifically modified *E. coli* strains or cell-free expression systems with low levels of metabolic conversion.

Currently, prior to main NMR data acquisition, target proteins are evaluated routinely as to their suitability for NMR studies by a relatively easy heteronuclear single quantum correlation experiment (^{15}N-^{1}H-HSQC). The aim of this step is to filter out improperly folded targets, for which doing further more complex, time-consuming, and expensive experiments would be pointless. Moreover, the protein sample should be reasonably stable at high concentrations during the length of time required for NMR measurements. A broad range of conditions (e.g., pH, protein concentration, temperature, ionic strength, type of buffer, stabilizing and solubilizing additives) is usually tested to optimize the solubility and stability of the target protein, to lower its tendency to aggregate, and to increase the quality of recorded NMR spectra. The drawback of this attitude is that solvent conditions optimal for NMR measurements might be different from the physiological conditions characteristic of the protein of interest.

To acquire NMR data, the sample is placed in a probe installed in the center of a magnetic field generated by an NMR spectrometer equipped with a superconducting magnet (generating a static magnetic field) and digital frequency synthesizers (generating an oscillatory magnetic field in a perpendicular direction), and the resulting free induction decay (FID) signals are detected and Fourier-transformed to produce the NMR spectrum.

The structures of small molecules can be determined by basic one-dimensional ^{1}H-NMR. However, the increasing molecular weight results in a linear increase in the number of NMR signals, reduced resolution, spectral overlap, and decreased signal-to-noise ratio, thus making the spectra interpretation exceedingly problematic. The determination of structures of larger molecules requires a multidimensional NMR experiment performed by varying additional experimental parameters (i.e., by applying specially designed pulse sequences of different radio-frequency fields with variable duration and time intervals). Consequently, peak resolution is greatly improved and additional structural data are provided compared to the one-dimensional experiment because of the capability of recording selective pair interactions among spins. There is a huge variety of differently tailored multidimensional homonuclear and heteronuclear NMR experiments, among which the most essential are the variants of COSY (correlated spectroscopy), which supplies J-coupling constants that carry information on through-bond electron-mediated spin-spin interactions, and NOESY (nuclear overhauser effect spectroscopy), which provides internuclear distances that can be used to construct the protein model.

Usually, a set of individual heteronuclear multidimensional NMR experiments is performed to provide a sufficient amount of data for calculation of the three-dimensional protein structure (Figure 4). The interpretation of NMR spectra is based on the analysis of a vast number of parameters, such as chemical shifts, NOEs, or J-coupling constants, and includes iteratively repeated rounds of resonance assignment, collection of conformational constraints to deduce structural information, and calculation of the structure. This procedure is extremely complicated and laborious and represents the major bottleneck in high-throughput structure determination by NMR spectroscopy since it sometimes still involves manual examination of individual peaks. However, major progress in the development of computational tools to

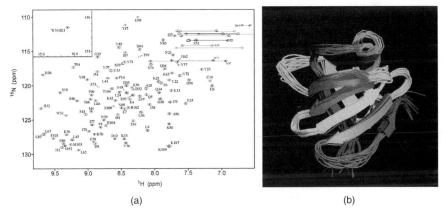

(a)   (b)

FIGURE 4 Determination of the three-dimensional structure of staphostatin A (*S. aureus* cysteine proteinase inhibitor) by NMR spectroscopy: (a) H^{15}N HSQC spectrum; (b) model of the three-dimensional structure. (Courtesy of G. Dubin.) (*See insert for color representation of figure.*)

facilitate automation of data processing and calculation procedures has been made recently, and further improvements are ongoing.

NMR structure calculation and refinement returns an assembly of conformers that represent the protein structure, all of which should be in agreement with the input experimental data. The quality of the model constructed is then evaluated and its coordinates are deposited in the PDB.

In general, the time needed for determination of the protein structure by NMR is substantially longer than that required by x-ray crystallography, given that diffraction-quality crystals are available. Several previously significant obstacles have been circumvented in recent years by break-through advancements in NMR hardware. They include the implementation of ultra-high-field magnets and cryoprobes to cool the electronic detecting systems of a spectrometer using ultracold helium gas, which improves the signal-to-noise ratio, dramatically increases measurement sensitivity, and reduces the time required for data collection. Large-scale NMR analysis was facilitated further by the introduction of sample changers and flow probes working in automated, serial mode. The main step forward in NMR methodologies was made recently by the development of *transverse relaxation-optimized spectroscopy* (TROSY). TROSY greatly improved the sensitivity and extended the size limitations of solution NMR spectroscopy from molecules of 20 to 30 kDa to, at present, 30 to 50 kDa. Moreover, spectra of proteins with molecular mass higher than 100 kDa have already been recorded, and it is expected that this limit will come up to about 500 kDa. Thus, a structural proteomics study by NMR spectroscopy might soon evolve from complementary to x-ray crystallography in the analysis of proteins that fail to crystallize, to the technique of first choice (e.g., for membrane protein structure determination or investigation of protein–protein interactions in macromolecular complexes).

The Protein Data Bank (PDB) was created in 1971 to use the increasing amounts of structural data in a more efficient way. Data on protein structure deposited in the PDB by means of the Internet include coordinates of atoms that make up the molecule, literature references, basic information on a molecule (such as the name, sequence, and organism of origin) and its tentative ligands, details about the method used to solve the structure, and some raw experimental data. Moreover, the nomenclature and data format are standardized and a number of tests are conducted to verify that the information introduced is correct (e.g., the proposed structural model is compared with experimental data and amino acid sequence). Adjustments to a common standard is necessary for efficient exchange of data, which is a prerequisite for further development of structural proteomics.

PROTEIN STRUCTURE PREDICTION

In parallel to development of experimental structure determination methods such as x-ray crystallography and NMR spectroscopy, much effort is being expended to advance computational methods of theoretical prediction of protein structures in silico, based on their amino acid sequences. In general, the issue of protein structure prediction is addressed either by comparative modeling, which scans the unknown sequence against protein databases, or by ab initio computing trials that neglect direct information contained in protein databases and are based exclusively on laws of quantum mechanics. Ab initio protein modeling is currently capable of providing accurate models for very small protein domains only, and comparative modeling remains the most powerful structure prediction approach. Nevertheless, ab initio modeling might be considered complementary to experimental structure determination as well as to comparative modeling since it might be used to predict structures of small protein portions that failed to be solved in other ways.

The first step in comparative protein modeling is fold assignment (or fold recognition), accomplished by searching databases to identify protein structures with sequences similar to the target sequence. This is usually done by multiple or pairwise sequence alignment, or if the basic sequence-to-sequence alignment does not recognize structurally characterized homologs of the protein being investigated, by *threading*, which scans the linear amino acid sequence of target protein against databases of known three-dimensional structures. The search returns the set of most compatible protein structures, which may serve as three-dimensional templates in modeling. Next, more detailed sequence–structure alignment is performed to find the most accurate fit between structurally equivalent residues in the target and template sequences. Subsequent model building is based initially on the alignment and the structures of the templates, while at a later stage it might require supplementary modeling of amino acid segments that differ substantially from the template (e.g., by ab initio methods). Finally, the quality of the model is tested and refined in iteratively repeated cycles of modeling and evaluation until the structure meets specific geometric and, most important, energetic criteria.

Understandably, the accuracy of structure predicted by comparative modeling corresponds to the level of sequence homology between a target protein and the template selected. Accordingly, models based on 40 to 50% sequence identity reach a quality that is comparable to low-resolution structures determined by x-ray crystallography and medium-resolution NMR structures. On the other hand, typically, high structural conservativity of protein active sites allows their modeling with relatively good accuracy despite the rest of the molecule, so that even templates with a poorer (down to 30%) overall sequence similarity to target protein may provide structures that are true enough to give some insight into protein function, and at least contribute to planning supplementary experimental work.

The effectiveness of the comparative modeling approach is derived from the limited number of distinct stable folds that can be adopted by a polypeptide chain. The most abundant folds, such as an eight-stranded α/β barrel of triosephosphate isomerase (TIM), may be found in as many as 20 individual protein families. The rapidly growing number of experimentally solved structures continuously enriches a source of suitable templates for comparative modeling while increasing the accuracy of the models generated. On the other hand, the massive amount of data resulting from the feasibility of sequencing of the entire genomes has already overwhelmed the processing power of experimental structure determination techniques. Thus, comparative modeling is a method that becomes not only a complement, but also an inevitably realistic prospect for structural proteomics research.

The automation of comparative modeling procedures is not yet highly advanced since human intervention is still required at many stages. However, the techniques used are developing rapidly, their progress being evaluated every second year through a CASP (critical assessment of techniques for protein structure prediction) contest. The task of participants is to predict the three-dimensional structure of a protein based on its amino acid sequence. The model proposed is then compared with the model resulting from experimental NMR or x-ray analysis. Consequently, the current potential of structure prediction methods is monitored on a regular basis, at the same time promoting both the development of computational tools and the exchange of knowledge, which are the most important CASP outcomes.

STRUCTURE-TO-FUNCTION ANALYSIS

Structural proteomics proceeds from gene sequence to the three-dimensional structure of a protein encoded by this gene, creating a new challenge to link pure structural information to the biochemical and biological function of the macromolecule. Structure-based functional study usually begins with finding how similar the protein of interest is to all other proteins whose structures (and often functions) are already known. Atomic coordinates of the target protein are submitted to a selected server using structural alignment techniques [e.g., DALI (distance alignment algorithm) or CE (combinatorial extension algorithm), and scanned against databases that structurally classify PDB entries, such as SCOP (structural classification of protein) or CATH (class, architecture, topology, homologous superfamily protein classification database)]. This returns the list of proteins with related overall folds.

The newly determined structure may belong to one of three groups: (1) proteins whose functions may be inferred from similarity of their structures to structures of proteins with already identified functions; (2) proteins whose structures have no significant homology to any other protein structure available in the databases, and (3) proteins that have structural homologs but where functions have not been unveiled.

Moreover, the structural similarity identified may cover either entire molecules or only their fragments. Then both the size of the shared structural similarity region and the exactness of overlap between the structures are of great importance for estimating how close the structure being analyzed is related to these found in the database and how much functional information might be inferred.

There are times, when structural similarity of a target structure to the functionally characterized protein in the database is so striking that the function of the protein of interest might be identified almost directly. In the absence of clear structural homology on a global level, it may still be possible to recognize specific three-dimensional substructures that hold the essence of protein biochemical function. The most often mentioned examples of such motifs are the helix–turn–helix (HTH) motif characteristic to DNA-binding proteins and the EF hand, which is responsible for calcium binding. The motif-based function prediction methods include construction of three-dimensional templates that capture amino acid residues important for the function by superimposing the functionally related structures, followed by assigning the novel protein to the template that is most alike. Yet another emergent approach is based on the assumption that intermolecular interactions between proteins (or between proteins and another molecules) occur because of the complementarity of their molecular surfaces. The surface of the protein of interest is modeled and analyzed with regard to features such as hydrophobicity, electrostatic potential, or the presence of either exposed protein–protein interaction sites or the clefts on the surface that may be accessed by a solvent or ligand. The most recent and powerful structure-based function prediction algorithms combine all the above-mentioned approaches and first search for proteins having the same, or a related, overall fold, then analyze the molecular surfaces, and finally, investigate a number of structural templates.

It is important to emphasize that the assignment of function to a protein based on its structure is a complex process that requires human intervention at many stages since one biochemical function may be associated with a variety of folds, while the same fold is often found in numerous, clearly distinct protein families that perform diverse functions. Therefore, to link protein structure and its biochemical function, it is always necessary to examine carefully the architecture of the active site of the investigated molecule.

Moreover, it should be noted that although knowledge of protein structure provides clues about its probable function, the physiological role of a protein may be highly complex and additional experimental data are usually required to understand it. On the other hand, information on protein structure and its tentative biochemical functions helps us to design further experiments (such as site-directed mutagenesis or synthesis of low-molecular-mass ligands) necessary to fully understand the biological role of that particular molecule.

LIMITATIONS AND PROSPECTS

Despite enormous progress in structural proteomics witnessed in recent years, this exciting field is still seriously limited by shortcomings in its technologies. The pace at which new protein structures are solved is much slower than that of genome sequencing, which urges the need to develop faster and more efficient methods of sample preparation, x-ray and NMR measurements, structure solving and refinement, comparative modeling, and data interpretation. Structural proteomics laboratories strive to automatize these tasks and, at the same time, try to find alternative ways of optimization of the entire procedure.

Efficient methods of crystallization of membrane proteins are still lacking. As a resultant, although most protein structures deposited in the PDB were solved by x-ray crystallography, only a few of them are those of membrane-bound macromolecules. At the same time, it is estimated that about 30% of all naturally occurring polypeptide chains are associated with the membranes, while G-protein-coupled receptors and ion channels, formed by integral membrane proteins, constitute 50 to 70% of potential human drug targets. Researchers try to face the growing need for structural and functional data about membrane proteins by improving the methods of their purification and crystallization, as well as by further developing NMR spectroscopy, as exemplified by recent development of the TROSY technique for the study of micelles.

Over the last few years, the major progress has been done toward coordination of activities in the field of structural proteomics. Since only about 30% of protein structures currently available in the PDB may serve as unique templates for comparative modeling, it is necessary to analyze all known amino acid sequences systematically in order to define protein families with 30 to 35% of sequence identity, the level considered as threshold for successful comparative modeling. Then the structure of at least one member protein of each such family should be determined experimentally in order to generate models of other.

Revealing new polypeptide chain folds will allow modeling of proteins, the structural data of which were not available before. Each model obtained in this way may facilitate the design of experiments that will provide information on the biological functions of proteins, such as site-directed mutagenesis, ligand-binding studies, measuring of enzymatic activity, and protein–protein interaction.

It is hoped that the results obtained by structural proteomics will greatly facilitate rational drug design. One idea is to apply a database of comparative models of protein structures for in silico analysis of ligand docking, with the use of libraries of low-molecular-mass organic compounds. This may lead to identification of potential therapeutics without resorting to traditional experimental screening. Yet another aspect is the perspective of identification of structures and functions of unknown proteins that as potential drug targets may be crucial for designing new treatment strategies. Based on an analysis of their binding sites, it will be possible to assess their usefulness as targets for particular small ligands. Understanding protein structures may also be helpful for the prediction of cross-interactions between low-molecular-mass compounds and proteins that were not even considered potential interacting partners. This will contribute to prevention of side effects of therapies, as certain compounds would thus still be excluded from the drug development process at the stage of preclinical studies.

Many current drugs were developed as a result of a joint effort of combinatorial chemistry and structure-based design. Common examples of drugs that were created based on protein structure–function studies include Zanamivir (anti-influenza drug), and Amprenavir and Nelfinavir (HIV protease inhibitors used for AIDS treatment). Highly efficient structure-based computer-aided drug design is still some distance away, but as more and more structural data are available, this aim will obviously become much closer.

RECOMMENDED READING

Acta Crystallographica, Section D, Biological Crystallography, 62 (Part 10), Oct. 2006.

Drenth J. *Principles of Protein X-Ray Crystallography.* Springer-Verlag, New York, 1999.

Petrey D., Honig B., Protein structure prediction: inroads to biology, 20. *Mol. Cell* , (2005) 811–819.

Structural Genomics Supplement. *Nat. Struct. Biol.* Nov. 2000.

Wütrich K. NMR studies of structure and function of biological macromolecules. Nobel lecture, Dec. 8, 2002.

Yee A., Gutmanas A., Arrowsmith C.H. Solution NMR in structural genomics. *Curr. Opin. Struct. Biol.* 16(2006) 611–617.

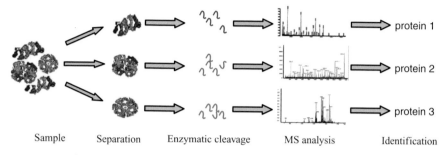

Sample Separation Enzymatic cleavage MS analysis Identification

Chapter 1, Figure 1 General strategy for protein identification.

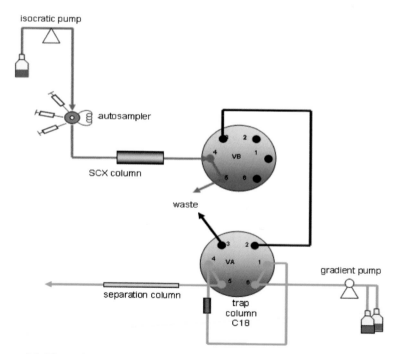

Chapter 3.2, Figure 1 Coupling scheme for SCX and RP columns with a small trap cartridge in between.

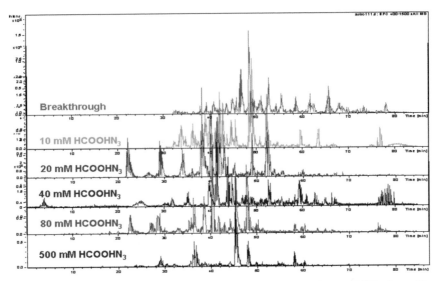

Chapter 3.2, Figure 2 Chromatograms presenting peptides eluted from an SCX column (first dimension) after subsequent separation on a reversed-phase capillary column. Concentrations of the salt segments (discontinuous gradient) are also shown.

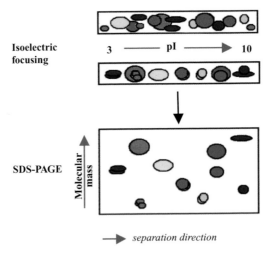

Chapter 4.2, Figure 1 Two-dimensional gel electrophoresis scheme: proteins sorted according to their net charge during an isoelectric focusing step are subsequently separated by molecular weight via SDS-PAGE, perpendicular to the IEF direction. Spots in color represent proteins of various shapes, sizes, and pI values.

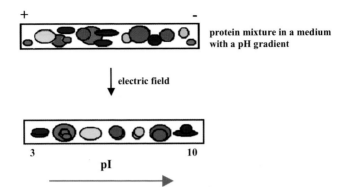

Chapter 4.2, Figure 4 Isoelectric focusing: migration of proteins in an electric field.

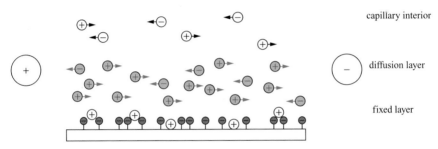

Chapter 5, Figure 3 The principle of electroosmotic flow. Silanol groups (Si–OH) abundant on the surface of the capillary wall dissociate with ejection of proton, resulting in the creation of a strongly negatively charged surface. The cations present in solution associate near the surface, the closest being attached strongest and immobilized near the wall. Cations situated behind the fixed layer in the diffussion layer, however, retain the ability to move along the capillary. After applying high voltage to the capillary ends, cations from the diffusion layer start to move toward the cathode, dragging buffer in the capillary with them.

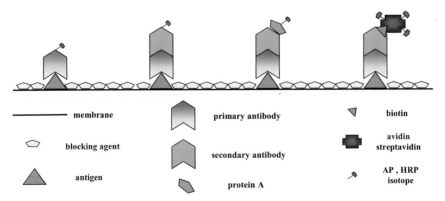

Chapter 6, Figure 3 Selected setups of protein immunodetection on membrane.

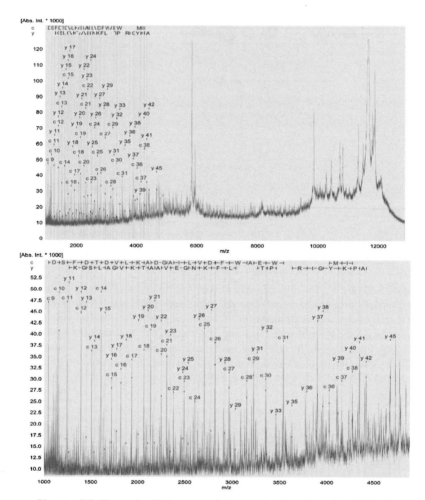

Chapter 7.2, Figure 2   ISD spectra of the protein thioredoxin (ca. 12 kDa).

Field of view: 18000.0 × 18000.0 μm²

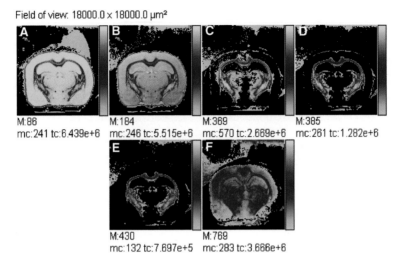

M:86
mc:241 tc:6.439e+6

M:184
mc:246 tc:5.515e+6

M:369
mc:570 tc:2.669e+6

M:385
mc:261 tc:1.282e+6

M:430
mc:132 tc:7.697e+5

M:769
mc:283 tc:3.666e+6

Chapter 7.3, Figure 4   Positive secondary ion images obtained from a rat brain section under the irradiation of Bi_3^+ primary ions. 256 × 256 pixels, pixel size 70 × 70 μm².

Field of view: 26000.0 × 26000.0 μm²

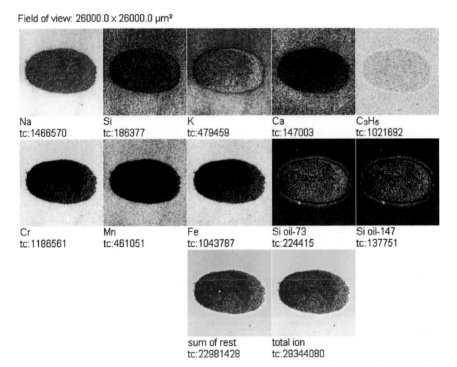

Na
tc:1466570

Si
tc:186377

K
tc:479459

Ca
tc:147003

C₃H₆
tc:1021692

Cr
tc:1186561

Mn
tc:461051

Fe
tc:1043787

Si oil-73
tc:224415

Si oil-147
tc:137751

sum of rest
tc:22981428

total ion
tc:29344080

Chapter 7.3, Figure 5   Image of a fingerprint taken from a stainless-steel sheet surface (26,000 × 26,000 μm²).

Field of view: 10000.0 × 10000.0 µm²

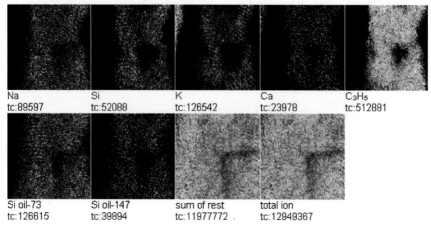

Na	Si	K	Ca	C₃H₅
tc:89597	tc:52088	tc:126542	tc:23978	tc:512881

Si oil-73	Si oil-147	sum of rest	total ion
tc:126615	tc:39894	tc:11977772 .	tc:12949367

Chapter 7.3, Figure 6 Image of a fingerprint taken from a paper surface (10,000 × 10,000 µm²).

Field of view: 10000.0 × 10000.0 µm²

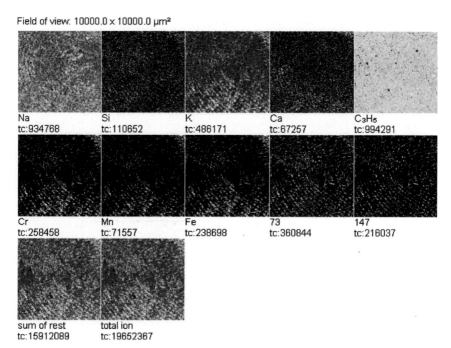

Na	Si	K	Ca	C₃H₅
tc:934768	tc:110652	tc:486171	tc:67257	tc:994291

Cr	Mn	Fe	73	147
tc:258458	tc:71557	tc:238698	tc:360844	tc:216037

sum of rest	total ion
tc:15912089	tc:19652367

Chapter 7.3, Figure 7 Image of a fingerprint taken from a stainless-steel sheet surface (10,000 × 10,000 µm²).

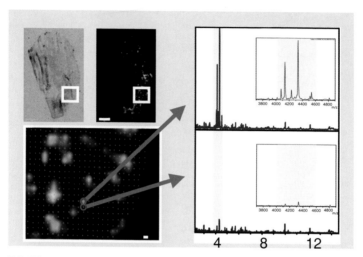

Chapter 7.3, Figure 8 MALDI imaging at high spatial resolution (50 μm). Beta-amyloid peptide plaques in Alzheimer mouse brain section.

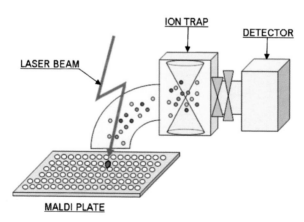

Chapter 7.3, Figure 9 AP- MALDI.

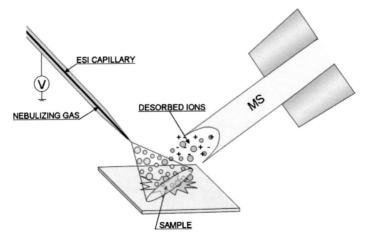

Chapter 7.3, Figure 10   DESI operational principle.

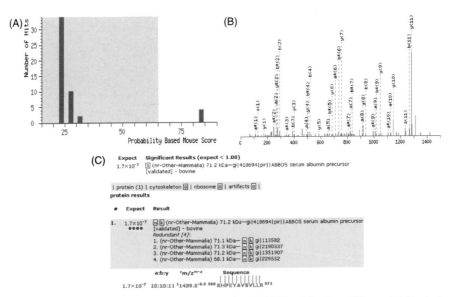

Chapter 7.5, Figure 9   Typical presentations of MASCOT and ProFound/Sonar. (A) Typical diagram for MASCOT results. The green shaded area represents the area where the confidence in the result is less than 95%. The red bars show the number of hits (height of the bar) as a function of the Mowse score. (B) Detailed result for an MS/MS spectrum with annotated peaks. (C) Results of a Sonar search using the same data set. Here, an expectation value (it tells us more or less how frequently we may expect a random event that gives a comparable result). The lower this value, the better. There is also an annotation of the fragment ion masses observed as lines above the sequence.

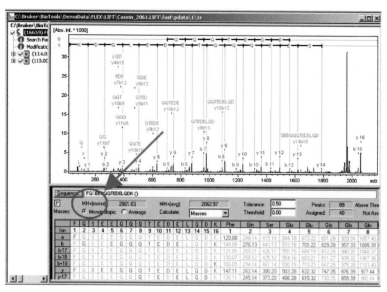

Chapter 7.5, Figure 10 Automatic detection and indication of peptide modifications in BioTools.

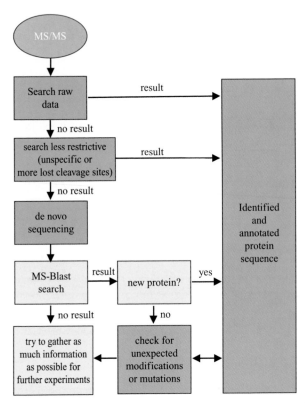

Chapter 7.5, Figure 11 Work flow generally recommended for the analysis of MS/MS data. The green boxes are tasks that can be performed using such software as BioTools, while the yellow boxes show tasks that involve other search engines or databases. Generally, this procedure should be followed stepwise from the easiest to the most difficult tasks. A comment to pacify the reader's mind: You will rarely encounter occasions where you will get 100% sequence coverage. (One of the authors (M.M.) has had only one case in eight years that involved a combination of MALDI and ESI as an ionization technique.) Therefore, having gaps or unclear regions in sequences is absolutely normal. This is especially the case when working with unsequenced genomes and having to rely on similar entries in the databases.

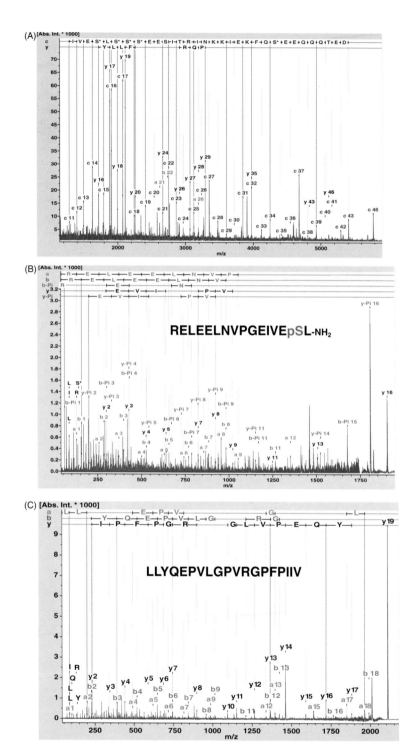

Chapter 7.5, Figure 13 T^3 spectra (ISD-LID) of β-casein in DHB: (A) ISD spectrum, T3 spectrum of c$_{16}$ (B) and of y$_{19}$ (C).

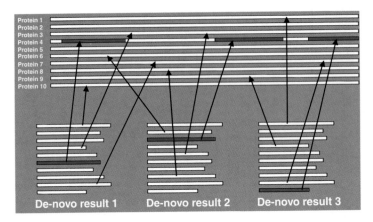

Chapter 7.5, Figure 14   Schematic view of the function of MS-BLAST.

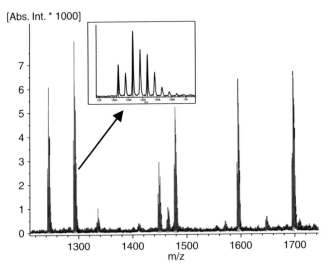

Chapter 7.6, Figure 10   MALDI mass spectrum of a proteolytic peptide mixture after enzymatic digestion in a mixture of ^{16}O and ^{18}O water. The inset shows a zoomed view to illustrate the isotopic peak pattern with the signal pairs resulting from incorporation of the isotopic mixture.

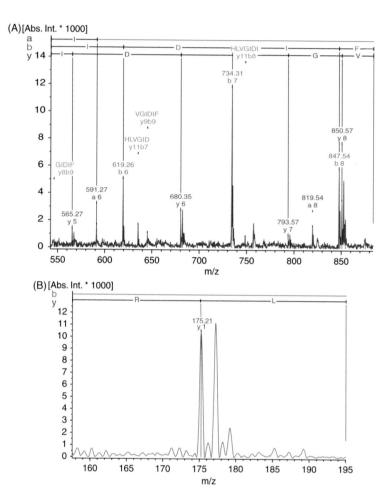

Chapter 7.6, Figure 11 (A) MALDI-MS/MS spectrum of a tryptic peptide. The annotated pairs always belong to y-type ions, while singlet peaks are either b-type, a-type, or internal fragment ions. (B) Zoomed view of the y^1 ion of a tryptic peptide with an arginine at the C-terminus. The doublet structure with 2 Da separation for the ^{16}O and the ^{18}O species is clearly visible. There is also a small third peak resulting from double incorporation of ^{18}O (from reentering of an already cleaved peptide into the active site of the enzyme and a second oxygen exchange by the enzyme). The longer the incubation time, the more pronounced the effect. It cannot be avoided completely.

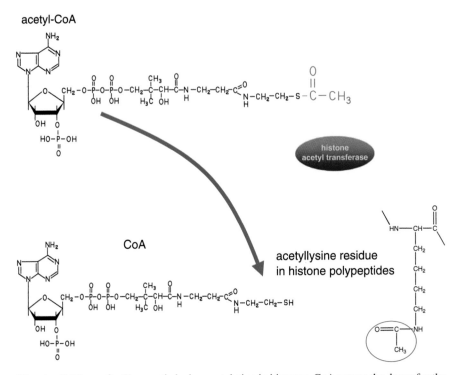

acetyl-CoA

CoA

histone acetyl transferase

acetyllysine residue in histone polypeptides

Chapter 8, Figure 3   Enzymatic lysine acetylation in histones. CoA acts as the donor for the acetyl moiety.

Cys • Ala • Aja • X

Cys • Ala • Ala • X

acceptor protein

farnesyl transferase

acceptor protein

farnesyl pyrophosphate

Chapter 8, Figure 5   Farnesylation at the C-terminus with the activated donor farnesyl prophosphate. The reaction occurs at Cys Ala Ala X sequences.

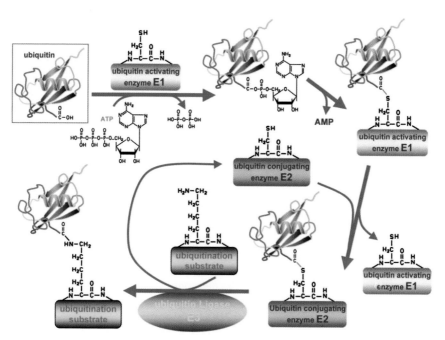

Chapter 8, Figure 6   Ubiquination on lysine of proteins destined to be degraded.

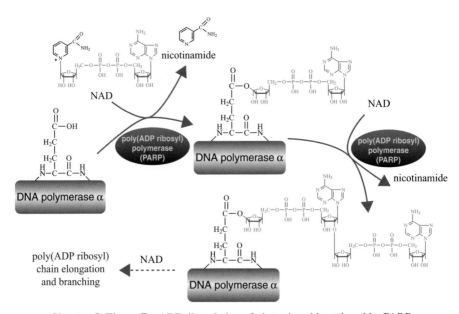

Chapter 8, Figure 7   ADP ribosylation of glutamic acid catalyzed by PARP.

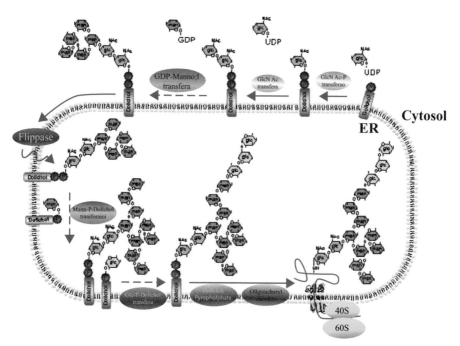

Chapter 8, Figure 8 N-Glycosylation pathway in the endoplasmatic reticulum.

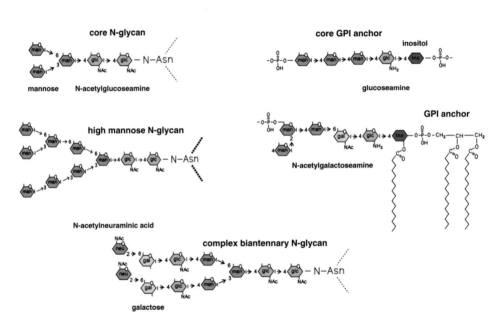

Chapter 8, Figure 9   Protein glycan modifications.

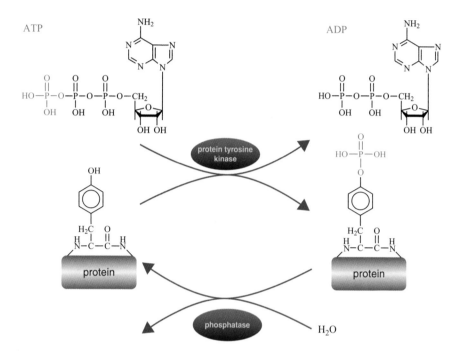

Chapter 8, Figure 10 Enzyme-catalyzed phosphorylation of tyrosine residues using ATP as the donor substrate.

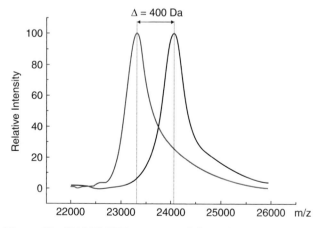

Chapter 8, Figure 13 MALDI-TOF spectrum of β-casein. This protein carries five O-phosphorylation sites. After removal of the phosphate by alkaline phophatase, the protein mass is decreased by 400 Da.

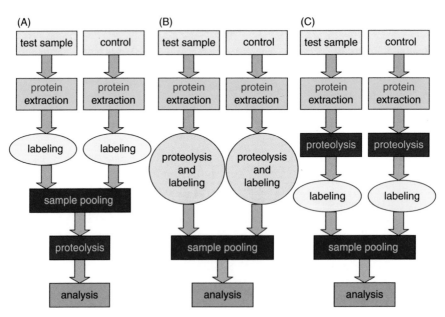

Chapter 8, Figure 14 Chelate complex formation of a phosphate group with immobilized metal ions. The chromatography can be performed with iminodiacetic acid or the tridentate nitrilotriacetic acid.

Chapter 10.1, Figure 4 General scheme describing procedures that involve isotope tags and mass spectrometry: (A) in vivo labeling; (B) in vitro labeling prior to proteolytic digestion; (C) in vitro labeling of peptide map.

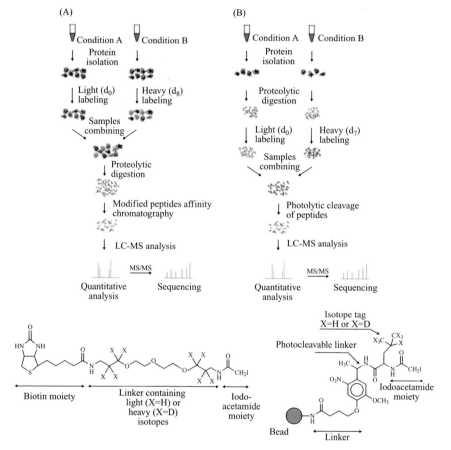

Chapter 10.1, Figure 5 Scheme of an ICAT experiment: (A) reaction in liquid; (B) solid-phase ICAT. The structures of the chemicals are shown at the bottom.

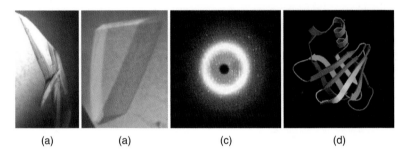

| (a) | (a) | (c) | (d) |

Chapter 10.2, Figure 3 Determination of the structure of staphostatin B (*Staphylococcus aureus* cysteine proteinase inhibitor) by x-ray crystallography: (a) crystal obtained in preliminary crystallization conditions; (b) crystal obtained after optimization of crystallization conditions; (c) diffraction pattern; (d) model of three-dimensional structure.

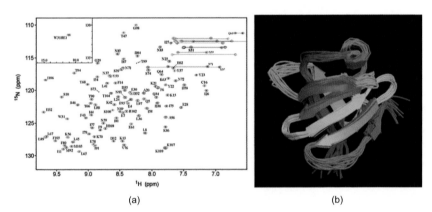

| (a) | (b) |

Chapter 10.2, Figure 4 Determination of the three-dimensional structure of staphostatin A (*S. aureus* cysteine proteinase inhibitor) by NMR spectroscopy: (a) $H^{15}N$ HSQC spectrum; (b) model of the three-dimensional structure.

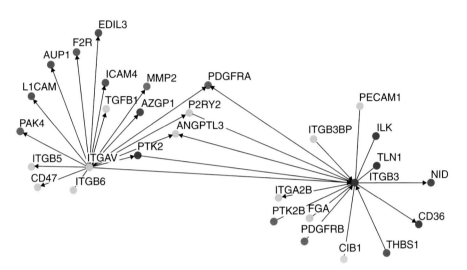

Chapter 10.3, Figure 7 Fragment of a human interaction map obtained by proteomic examination of prostate cancer cell lines.

Matched peptides shown in **Bold Red**

```
  1 MRFSCLALLP GVALLLASAL LASASDVLEL TDENFESRVS DTGSAGLMLV
 51 EFFAPWCGHC KRLAPEYEAA ATRLKGIVPL AKVDCTANTN TCNKYGVSGY
101 PTLKIFRDGE EAGAYDGPRT ADGIVSHLKK QAGPASVPLR TEDEFKKFIS
151 DKDASVVGFF RDLFSDGHSE FLKAASNLRD NYRFAHTNVE SLVKEYDDNG
201 EGITIFRPLH LANKFEDKIV AYTEKKMTSG KIKKFIQESI FGLCPHMTED
251 NKDLIQGKDL LTAYYDVDYE KNTKGSNYWR NRVMMVAKTF LDAGHKLNFA
301 VASRKTFSHE LSDFGLESTT GEIPVVAIRT AKGEKFVMQE EFSRDGKALE
351 RFLQEYFDGN LKRYLKSEPI PETNEGPVKV VVAESFDDIV NAEDKDVLIE
401 FYAPWCGHCK NLEPKYKELG EKLSKDPNIV IAKMDATAND VPSPYEVKGF
451 PTIYFSPANK KLTPKKYEGG RELNDFISYL QREATNPPII QEEKPKKKKK
501 AQEDL
```

Appendix A, Figure 4 Sequence of protein identified; the red color denotes peptide fragments detected along the mass spectrum.

Score: 8.55, 39 matching peptides: P11598 (PDA3_RAT)pI: 5.78, Mw: 54239.39

CHAIN 1: Protein disulfide isomerase A3 - Rattus norvegicus (Rat).

| GlycoMod | FindMod | FindPept | PeptideMass | BioGraph |

user mass	matching mass	Δmass (Dalton)	#MC	modification	position	peptide
948.5767	948.5118	-0.0648	1		281-288	NRVMMVAK
997.2895	997.5101	0.1206	0		153-161	DASVVGFFR
1172.4668	1172.5405	0.0737	0		336-344	FVMQEEFSR
1179.4675	1179.5865	0.119	1		174-183	AASNLRDNYR
1188.4446	1188.5354	0.0908	0	MSO: 338	336-344	FVMQEEFSR
1191.6078	1191.6004	-0.0073	0		63-73	LAPEYEAAATR
1194.6338	1194.6729	0.0371	2		416-425	YKELGEKLSK
1236.4681	1236.5127	0.0446	0		108-119	DGEEAGAYDGPR
1244.5771	1244.6634	0.0863	0		184-194	FAHTNVESLVK
1284.1189	1283.5255	-0.5833	0		83-94	VDCTANTNTCNK
1341.604	1341.6838	0.0798	0		449-460	OFPTIYFSPANK
1342.621	1342.6889	0.0679	1		215-225	FEDKIVAYTEK
1373.6019	1373.6736	0.0717	0		352-362	FLQEYFDGNLK
1394.597	1394.6587	0.0617	0		162-173	DLFSDGHSEFLK
1395.6305	1395.6876	0.0571	2		272-282	NTKOSNYWRNR
1396.5647	1396.6954	0.1307	0		367-379	SEFIPETNEOPVK
1397.6177	1397.5784	-0.0392	0	2xCys_CAM	83-94	VDCTANTNTCNK
1397.6177	1397.706	0.0883	0		472-482	ELNDFISYLQR
1469.6524	1469.7787	0.1263	1		449-461	OFPTIYFSPANKK
1488.6838	1488.6787	-0.005	1	MSO: 338	336-347	FVMQEEFSRDGK
1529.6964	1529.7347	0.0783	1		352-363	FLQEYFDONLKR
1547.2348	1547.8073	0.5725	1		283-296	VMMVAKTFLDAOHK
1587.7407	1583.8165	0.0758	1		148-161	FISDKDASVVGFFR
1588.742	1588.8805	0.1385	2		62-75	RLAPEYEAAATRLK
1632.6891	1632.7472	0.0581	0	MSO: 434	434-448	MDATANDVPSFYEVK
1632.6891	1632.7663	0.0772	1		105-119	IFRDGEEAGAYDGPR
1653.6762	1653.7602	0.084	0		25-38	SDVLELTDENFESR
1715.8715	1715.9115	0.04	2		147-161	KFISDKDASVVGFFR
1721.7854	1721.9432	0.1578	1		483-497	EATNPPIIQEEKPKK
1744.9069	1744.8864	-0.0204	1		131-146	QAGPASVPLRTEDEF K
1746.8636	1746.9286	0.065	1		289-304	TFLDAGHKLNFAVAS R
1749.9178	1749.8541	-0.0636	0		380-395	VVVAESFDDIVNAED K
1780.9344	1780.8185	-0.1158	0		396-410	DVLIEFYAPWCOHCK
1792.7308	1792.8977	0.1669	1		180-194	DNYRFAHTNVESLVK
1801.0591	1800.9378	-0.1212	1		364-379	YLKSEFIPETNEOPV K
1872.9192	1872.9814	0.0622	2		130-146	KQAGPASVPLRTEDE FK
1872.9192	1872.9614	0.0622	2		131-147	QAGPASVPLRTEDEF KK
2302.0771	2302.1462	0.0691	0		195-214	EYDDNOBOITIFRPL HLANK
2605.2548	2605.3032	0.0484	2		253-274	DLIQOKDLLTAYYDV DYEKNTK
2605.2548	2605.3144	0.0596	0		306-329	TFSHELSDFOLESTT OEIPVVAIR
2717.7311	2718.3158	0.5847	2		95-119	YOVSOYPTLKIFRDO EEAGAYDGPR
2733.359	2733.4093	0.0503	1		305-329	KTFSHELSDFOLEST TOEIPVVAIR
2820.2194	2820.3695	0.1501	1	1xCys_CAM	235-258	FIQESIFOLCPHMTE DNKDLIQOK

ΔMw: 24239.4 Da (30.3%)
81.3% of sequence covered:

```
         1        11        21        31        41        51
         |         |         |         |         |         |
        61 bDLAPEYEAA AIRLEgtoyl aWUCTANTN TCNMTOVEGT PTLKIFEDGE EAGAYDGPPs    60
       121 adgivshlMG QAGPASVPLR TEDEFEDFIS DRDASVVGFF RDLFSDGHKE FLKAASNLRD   120
       161 MYRFAHTNVS SLVKEYDDNG KGITIFRPLH LAHLFEDKIV AYTEKutng krkkVTIQESI   180
       241 FGLCPHMTED MDKLIQKDL LTAYYDVDYE RNTKGSNYWR MRNVMMVAKTF LDAGHKLNFA    300
       301 VSGNKTFSHE LSD-FQLESTT OEIPVVAIR= akpeKFVMEK EPSRDGKMals rFLQEYFDGN   360
       361 LKBYLKSEFI PETNEOPVNV VVAESFDDIV NAEKDQVLIR FYAPWCGHCK nlepkVNELG    420
       471 ESLKGdpniv iakHDATAND VPSFTEVKKP PIIYFSPANK Kltpkbyegg rELNDFISYL    480
       481 GREATNPPII QEEKPKKkkk aqedl                                         480
```

Appendix A, Figure 7 Results of the search in the ExPASy database using the PeptIdent program. The sequence of the identified protein is given below. Note the accuracy of the calculations and measurements (four decimal digits), which cannot be obtained during low-resolution analysis.

10.3

FUNCTIONAL PROTEOMICS

TOMASZ DYLAG, ANNA DRABIK, AND ADAM MOSZCZYNSKI

Functional proteomics focuses on systems of biochemical reactions that govern the functioning of single cells, tissues, and whole organisms (Figure 1). In particular, it encompasses:

- Identification and characterization of proteins that act in physiological processes
- Characterization of interactions of proteins with other proteins, nucleic acids, and low-molecular-mass compounds
- Determination of composition and function of macromolecular complexes
- Investigation of interdependencies between various biochemical pathways and uncovering their regulatory mechanisms

The genome may be treated as "static," whereas the proteome is subject to dynamic changes and is dependent not only on the cell type but also on the conditions in which the cell is found at a given time. The function of a protein is determined by its amino acid sequence but is also tuned by posttranslational modifications. Additionally, usually only upon formation of multimolecular complexes are single proteins endowed with particular activities. For all these reasons, the investigation of functions of a protein separated from its natural surrounding may lead to false conclusions, while the conditions present in a living organism are difficult to reproduce in a typical in vitro experiment. Thus, another important goal of functional proteomics is to develop methods to assess the functions of particular proteins in conditions most closely resembling natural conditions.

Understanding protein functions has significant scientific value and can be implemented relatively quickly, in particular in drug design, as malfunctioning of a single

Proteomics: Introduction to Methods and Applications, Edited by Agnieszka Kraj and Jerzy Silberring
Copyright © 2008 John Wiley & Sons, Inc.

FIGURE 1   Area of interest of functional proteomics.

protein may cause disease. Although functional proteomics is nominally focused on the function of proteins, it often relies on methods innate to molecular biology or involves genetic intervention, and examination of the functions of proteins provides insight into the roles of corresponding genes. Therefore, in recent years the area of interest of functional proteomics has been more and more overlapping with that of functional genomics.

HOW TO STUDY PROTEIN FUNCTION

Proteins exert their roles in living organisms by interacting with other proteins, nucleic acids, lipids, saccharides, and other classes of compounds; therefore, the functions of proteins can be examined by analyzing their interactions. As a first step to understanding the role an unknown protein, its subcellular localization should be probed, as the two are often strongly correlated. This can be achieved by classical methods of fractionation of organelles, followed by two-dimensional gel electrophoresis and mass spectral analysis. If specific fluorescently labeled antibodies against the protein of interest are available, they can be used to detect it over different tissues and cell compartments by microscopy (the approach known as *immunohistochemistry*). When the protein sequence is known, its localization could be predicted in silico by checking for the presence of specific fragments such as signal peptides or nuclear localization sequences.

Protein function can be understood by studying its interactions with proteins already known, since interacting partners are often elements of the same pathway. Most biochemical methods of functional studies are based on the use of molecules of various types that interact with the protein of interest, and on subsequent characterization of these interactions. More sophisticated methods of investigation of protein functions are based on their in vivo inhibition. If a particular process is perturbed after the expression of a single gene has been blocked, the protein products of this gene (i.e., the proteins it encodes) are probably indispensable for the process to occur. Protein biosynthesis can be transiently inhibited by gene-silencing techniques, and further information can be obtained from experiments with transgenic animals. The roles of proteins in the most complex reactions of organisms can thereby be assessed. Protein functions may also be predicted in silico, based on the analysis of their three-dimensional structures, as described in Chapter 10.2.

AUTOMATION AND HIGH-THROUGHPUT RESEARCH

Despite their numerous advantages, methods of traditional biochemistry usually allow for investigating interactions between only single proteins at a time. In recent years, automated high-throughput research has been gaining popularity, as it enables fast and efficient study of the activity of many proteins within the framework of one experiment. Automation offers the following advantages:

- Relatively short time necessary to conduct an experiment
- Low sample consumption
- Limited amounts of reagents consumed and, consequently, wastes
- Minimal danger of contamination
- Improved reproducibility of results
- Minimized risk of human errors

For example, in a "classical" proteomic experiment based on two-dimensional electrophoresis, several stages can be automated, such as pipetting of samples of minute volumes, gel casting, electrophoretic separation, image analysis, spot excision, protein digestion and derivatization, chromatographic separation, mass spectral detection, data gathering, and database search. Together with automation, there is a trend for miniaturization, easily observed in array technologies. One of its major advantages is limitation of the amounts of precious biological material required (e.g., blood sample).

In fact, for advanced proteome-wide experiments to become feasible, automation and high throughput are indispensable. In functional proteomics, these requirements are met, among others, by peptide and cDNA libraries, and by protein, peptide, and DNA arrays. It is still the role of humans, though, to ensure that the data obtained by high-throughput techniques are correct.

High-throughput experiments often prompt novel research goals, such as investigation of new binding partners to proteins known previously. Therefore, the results obtained in this way are usually a starting point for further in-depth studies whose aim is to gain a thorough understanding of the role of a particular protein in a given physiological or pathological process. Such investigations are, however, often conducted with more traditional methods of biochemistry or molecular biology.

PROTEIN AND PEPTIDE ARRAYS

Arrays are a modern tool of proteomics used to search for particular peptides and proteins, and to analyze their activities. Protein and peptide arrays have been developed in analogy with DNA arrays that had been known beforehand. They allow us to conduct complete experiments in a high-throughput design and offer the added value of precisely controlling experimental conditions while providing a wealth of information. The production of arrays is automated, similar to sample deposition and detection, while miniaturization helps to limit the amount of sample and chemicals required. Protein and peptide arrays facilitate the use of proteomic methods in diagnostics, replacing complicated chromatographic or electrophoretic analyses. They are used both in basic research and by the industry for drug discovery purposes.

Peptide Arrays

Peptide arrays contain a number of peptides differing in their sequences that are covalently bound to the surface of a solid polymeric support.

The general idea behind the generation of a peptide array is similar to that of traditional solid-phase peptide synthesis (i.e., chain elongation from carboxy to amino terminus). Peptides are usually synthesized "on site" by the SPOT technique, in which the spots on the flat surface of support are functionalized with C-terminal amino acid derivatives. Droplets of solutions of different activated amino acid derivatives[1] are deposited at selected positions on the support surface for a time sufficient for the coupling to be efficient. Afterward, the liquid reaction mixture is washed away and the coupling cycle is repeated with different amino acids toward the N-terminus. The sequence of peptides synthesized in each position of the support is known.

The peptide molecule and the support are usually linked by a spacer that ensures enhanced conformational freedom and facilitates access to immobilized peptides of components of the sample analyzed. Since each spot on the surface contains a single peptide only (in many copies, though), functional tests may be conducted for each peptide simultaneously without the risk of interaction occurring between them or competing for binding partners. A peptide array offers the additional asset of high local peptide concentration, thanks to which it can interact simultaneously with many copies of partner molecules.

Examples of the use of peptide arrays include studying substrate specificity of proteolytic enzymes, epitope mapping, and analysis of interactions of protein fragments or domains with other proteins or DNA (Figure 2). For example, the array components may be designed by in silico proteolytic digestion of proteins, so that

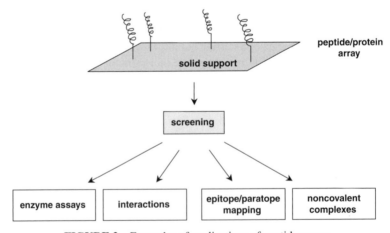

FIGURE 2 Examples of applications of peptide arrays.

---

[1] Amino acid derivatives, but not unmodified amino acids themselves, are used, as the latter are chemically inert toward other and must first be modified chemically to obtain reactive species. The side-chain functional groups of some amino acids are usually protected to prevent them from entering unwanted reactions.

conclusions on the activity of the full protein molecules can be drawn without having to obtain proteins by DNA recombination or by isolation from natural sources. Such peptide arrays are characterized by uniform physicochemical properties and allow us to eliminate a drawback of protein arrays, which stems from the vulnerability of proteins to the outside surroundings.

The use of peptide arrays in the determination of substrate specificity can be explained taking as an example an array composed of a series of potential peptidergic substrates of proteolytic enzymes. The spacer between the support and the peptide sequence includes a fluorophore, while the fluorescence quenching groups are attached to the peptide N-termini. The solution of the enzyme being investigated is deposited on the array surface, but the enzyme will cleave only specific amino acid sequences. Upon cleavage, the quenching group is removed, which causes the fluorescence to increase. Measurement of fluorescence allows us to determine in which array positions the peptides were cleaved. Knowing the sequence of peptides in each array position allows for the readout of sequences that were susceptible to cleavage.

Arrays designed to study the activity of particular types of proteins are commercially available. For example, one can use an array of peptides containing potential phosphorylation sites and amino acid sequences recognized by kinases (i.e., enzymes that catalyze coupling of the phosphate residue). The position of each single peptide in the array and its sequence are predetermined. During incubation of the array with the kinase and its labeled substrate, $[\gamma\text{-}^{32}\text{P}]\text{ATP}$, the sequences recognized by the enzyme are phosphorylated with radioactive phosphatic moieties, which enables their subsequent detection by autoradiography. Careful analysis allows us to elucidate the substrate specificity of the particular kinase.

Protein Arrays

Protein arrays consist of a solid support with immobilized protein molecules. The support may have different characteristics, depending on the type of array. Basically, it may be a glass plate, membrane, beads, or a more traditional microtiter plate. There are various methods of preparation and use of protein arrays. Analytical (capture) arrays specifically bind certain sample components since their surface is covered with relevant antigens, receptors, or DNA-binding proteins. After binding partner molecules, all others are washed away, so that only specifically interacting ligands are left to be analyzed. The idea of analytical arrays is thus similar to that of affinity chromatography. They can be applied for comparison of samples corresponding to health and disease in order to discover biomarkers of pathology (protein expression profiling), or to check for the presence of particular proteins in body fluids or tissue extracts. Depending on their design, analytical arrays can be used to screen a sample for the presence of different proteins simultaneously, or to screen many samples in parallel to check for the same biomarker. Their main features include high sensitivity and ease of application.

For investigation of protein–protein interactions, functional arrays are used that contain several thousands of different proteins. They are employed to assess interactions of proteins with other proteins, DNA, or low-molecular-mass ligands; in particular, the interactions investigated may encompass those between receptor and ligand (or, specifically, drug), or between the enzyme and its substrate or inhibitor. A dedicated array makes possible complete analysis of the interactions of several versions of one protein, including its domains, mutants, or alternative splicing variant, with another ligand of choice. The arrays can also represent fragments of genomes of one organism (e.g., yeast).

Protein molecules to form the array are usually obtained from an expression library, often using high-throughput protein expression and purification methodology. A variety of expression systems are available, and a careful choice is necessary to obtain a representative set of proteins (see Chapter 10.2 for more information). Proteins are deposited on the surface by robotic systems able to deliver only minute volumes onto predefined positions, yielding densities of up to several hundred thousands of spots per square centimeter. They are immobilized by covalent interactions through the chemical affinity of modified protein molecules to functionalized support (e.g., based on biotin–avidin interaction) or by means of nonspecific adsorption.

The method of detection is dependent on the array type and on the nature of interactions under investigation. Most often it is based on fluorescence measurement using either labeled antibodies or nonlabeled ligand-specific (primary) antibodies and fluorescently labeled secondary antibodies. If sample components can be labeled in advance, the fluorescence of ligands retained by the array can be measured. Signal intensity is then proportional to the amount of bound molecules, which allows for quantitative conclusions. Other typical detection systems are based on the use of MALDI mass spectrometry, atomic force microscopy, or surface plasmon resonance.

Possible difficulties of application of protein arrays stem from the sometimes troublesome task of expression and purification of huge sets of proteins. Another risk is connected with the vulnerability of proteins to outside conditions. Immobilizing a protein on a two-dimensional surface can disintegrate its three-dimensional structure and render the protein partially or even fully inactive. Another nontrivial task is to immobilize the molecules in such a way that their binding sites or domains can be accessed easily by interacting partners. For antibody-based arrays, the problem of cross-reactivity cannot be neglected. Regardless of the method used, however, the method of binding proteins to the surface must ensure reproducibility of results in numerous high-throughput experiments. For all these reasons, arrays should be validated carefully before they are used in real-life studies.

PROTEINS IN MACROMOLECULAR COMPLEXES

In cells, proteins rarely carry out their functions in isolation. Rather, cascades of reactions are regulated by conglomerates of various macromolecules, sometimes of differing chemical nature (e.g., DNA binding proteins, enzymes, cofactors). Proteins that play roles in a particular process may be organized into macromolecular assemblies, as is the case for the pyruvate dehydrogenase complex. Its molecular mass approximates 5 MDa, and it contains three different enzymes that transform pyruvate into acetyl-coenzyme A in the course of several subsequent reactions. Therefore, the formation of a macromolecular complex suggests that its components interact with each other in a stable manner and take part in regulation of the same pathway. These interactions can be studied after isolation of intact complexes, which is often a difficult and tedious task.

The methods of choice to study protein complexes include mass spectrometry (described extensively in other chapters) and antibody-based techniques. An example of the latter is *coimmunoprecipitation*, by which one can isolate complexes from multicomponent mixtures such as cell lysates, tissue homogenates, or body fluids. In this approach, one of the complex member proteins (thus serving as an antigen) is recognized by a specific antibody. When the antigen–antibody complex is formed, it is captured by protein A immobilized on insoluble (e.g., agarose or sepharose) beads. Since the interacting molecules are bound to solid support, all unbound and presumably undesired sample components can be washed away. The complex is then eluted by, for example, decreased pH or denaturation, so that all interacting proteins can be analyzed by other methods, such as gel electrophoresis or mass spectrometry. For example, the complex is denatured to release single proteins that are subsequently separated by gel electrophoresis and digested by trypsin in order to identify interacting partners by LC-MS/MS. The choice of capturing protein (e.g., protein A or protein G) is dependent on the species used to raise the antibody.

A method very similar to coimmunoprecipitation is the *pull-down assay*, which differs only in that an immobilized bait protein is used to bind partner molecules instead of antibody recognizing the antigen, which in turn binds other proteins. Still, this bait is tagged by a molecule specifically recognized by an immobilized partner ligand (e.g. 6xHis tag is bound by immobilized Ni^{2+} ions).

Yet another method that has gained popularity is *tandem affinity purification* (TAP). To study interactions of a particular protein (e.g., one recruited by comparative proteomics techniques as a potential biomarker), DNA recombination must first be employed to obtain a construct with the additional C-terminal TAP tag composed of a calmodulin-binding peptide, an amino acid chain recognized by tobacco etch virus protease, and a fragment of protein A. This recombinant protein serves as a bait for other natural proteins with which it might interact to form a complex. After incubating the construct with the sample of choice (e.g., a cell lysate), the newly formed protein assembly is isolated by affinity chromatography against immunoglobulin G, which recognizes the protein A sequence. Bound complexes are separated from the support upon proteolytic cleavage by tobacco etch virus protease. Subsequently, the complex released is bound to immobilized calmodulin so that the enzyme molecules and other impurities can be washed away. The purified complex is then eluted and subjected to various additional steps of analysis. A potential drawback of TAP is the risk of loss of activity of the protein of interest caused by incorporation of the tag.

The procedures described above often fail if very weak and transient interactions are studied, since such protein assemblies are likely to dissociate during isolation procedure.

PHAGE DISPLAY

Phage display is an important method for high-throughput investigation of inter-actions of proteins with other peptides and proteins, including but not limited to hormones, antibodies, and proteolytic enzymes. This technique makes use of bacte-riophages (i.e., viruses that infect bacterial cells). In the initial phase of the experiment, phages receive extra genes that code for additional proteins, yet unknown ligands for our known desired target whose interactions we are studying. The genes are designed such that once expressed, their products are connected to the phage coat protein so that the new fusion proteins are presented on the outer phage surface. The phages are then incubated with a selected target protein immobilized on a petri dish (a process known as *biopanning*). During incubation, complexes of ligands are formed with the target proteins; thus, bound phages remain on the dish, while nonbound phages can be washed away. In the next stage, the genome of the phage selected is sequenced to determine the sequence of the newly recruited ligand protein. Then, once the sequence of the protein selected is known, it is cloned for further study (Figure 3).

A phage display library may be composed of billions of different ligands. It may contain, for example, substrates of proteolytic enzymes or sets of mutants of a protein designed such that mutations cover only surface-exposed amino acids likely to account for interactions, or selected elements of protein secondary structure such as α-helices or zinc fingers. Phage display libraries are also employed to select ligands for isolation of newly discovered proteins from tissue extracts by affinity chromatography, or to select epitopes recognized by antibodies. In addition to libraries coding for precisely designed sets of peptides or proteins, random peptide libraries are also used.

Phage display is one of the most efficient methods of lead structure selection; therefore, it is commonly used for drug discovery. After the initial selection of lead, a secondary library may be designed in order to find a ligand fulfilling very tight criteria. Apart from the approach described above, phage libraries can be tested against proteins present on the surface of living cells. The process of designing and

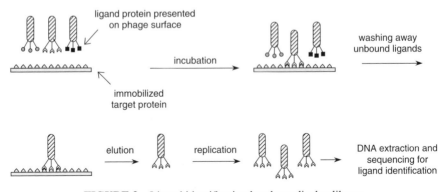

FIGURE 3   Ligand identification by phage display library.

testing a phage display library takes much less than traditional chemical screening. Phages can easily be amplified, so once created, a library can be used for many tests.

YEAST TWO-HYBRID SYSTEM

Yeast two-hybrid screening (also known as Y2H) is a powerful method of identifying binding partners to a known protein. It is based on the use of transcription factors (i.e., proteins that regulate transcription by interacting with specific DNA fragments); therefore, the process takes place in the yeast cell nucleus. A transcription factor comprises a DNA-binding and DNA-activation domain, and transcription can be initiated only when the two domains interact. Thus, in a yeast two-hybrid study, two proteins are constructed by DNA recombination: the bait (target, a known protein whose ligands we want to discover) connected to the DNA-binding domain, and the prey (test protein), connected to the transcription activation domain. Prey proteins are usually encoded in a cDNA or random DNA library, and plasmids with genes coding for the proteins investigated are transferred into yeast for expression.

If bait and prey proteins are to interact, they approach one another, and at the same time, the transcription factor domains coupled to them can interact as well. The transcription factor is designed so as to initiate the expression of a reporter gene that had been inserted into the yeast genome beforehand. This reporter gene codes for proteins necessary for the growth of cells on particular culture media (e.g., deprived of some nutrients), or for enzymes such as β-galactosidase, in the presence of which the yeast colony incubated with chromogenic X-Gal substrate turns blue. The expression of the reporter gene demonstrates that interaction occurs between two proteins (Figure 4). The DNA of selected colonies is sequenced for identification of the prey protein that interacts with the bait.

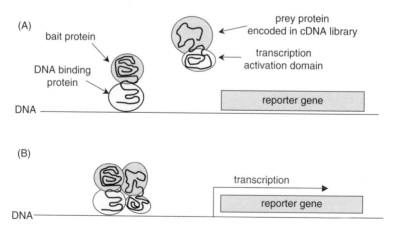

FIGURE 4 Yeast two-hybrid system: The bait is fused to the DNA-binding domain; the prey is connected to the activating domain of the transcription factor. (A) Bait and prey do not interact, so no transcription follows; (B) bait and prey do interact, so the DNA-binding domain approaches the activation domain to initiate transcription of the reporter gene.

The Y2H system is an efficient method of high-throughput investigation of protein–protein interactions. The bait proteins encoded in a library may represent, for example, components of a certain tissue or organism. By careful mutation of plasmid sequences, it is possible to construct a library of proteins differing only by selected amino acid residues in order to find those crucial for interactions. The throughput may be increased even more by testing a library of baits against a library of preys. The extent of transcription activation (thus also of the expression of the reporter gene) is dependent on the strength of protein–protein interaction; therefore, colorimetric assessment of β-galactosidase activity allows for quantitative comparison of interactions of the protein selected with different ligands. In principle, the yeast system enables us to study human proteins, since proteins coupled to transcription factor domains can also originate from other organisms. Y2H has been used successfully to construct proteome-wide interaction maps of several organisms, such as *Helicobacter pylori*, yeast, *Drosophila,* and humans.

Yeast two-hybrid assay as presented above may be conducted in other technical settings as well (e.g., a yeast one hybrid system for screening of protein–DNA interaction, or tested in *E. coli* as the host organism). One potential drawback of this and other methods based on the use of fusion proteins stems from the risk of their incorrect folding, which could hamper the access of ligands to sites crucial for interaction and hamper protein activity. For this and other reasons, due to the potential risk of false results or artifacts, data obtained by a yeast two-hybrid system should also be verified by other techniques, such as coimmunoprecipitation or pull-down.

GENE-SILENCING STRATEGIES

By blocking the activity of a particular protein, we can identify processes in which it plays significant roles. The best time to switch off the functioning of proteins is before they are formed (i.e., at the level of translation). This is also due to the fact that each mRNA molecule gives rise to many protein copies. The techniques described here are sometimes referred to as *knockdown*.

Antisense Oligonucleotides

Antisense oligonucleotides (AOs) are single-stranded oligonucleotides (DNA or RNA), typically 15 to 25 nucleotides, whose sequence is complementary to mRNA that results from gene transcription (Figure 5). The AOs have the power to block the biosynthesis of a particular protein without influencing the expression of other genes.

AOs form duplexes with mRNA transcripts. Upon formation of such a complex, mRNA becomes vulnerable to the activity of endogenous ribonuclease H, which cleaves it. The AO itself remains intact, so that after being released from the complex, it can bind another mRNA molecule and earmark it for destruction. Other possible mechanisms of action of AOs include formation of an inactive ribosome–mRNA complex or steric hindering of the access of a ribosome to a specific site in the mRNA chain (Figure 6).

```
5' ...AGTCCTAGCCTTAGCCATGCG... 3'      sense DNA strand
3' ...TCAGGATCGGAATCGGTACGC... 5'      antisense DNA strand
```

transcription of template DNA strand

```
5' ...AGUCCUAGCCUUAGCCAUGCG... 3'      sense mRNA
   3' TCAGGATCGGAATCGGTACGC   5'       antisense oligonucleotide
```

FIGURE 5   Sequence of sense mRNA fragment and its antisense oligonucleotide.

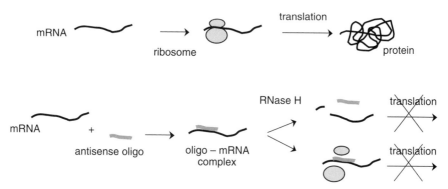

FIGURE 6   Transcription inhibition by antisense oligonucleotides. The mRNA strand incorporated in the AO–mRNA duplex is susceptible to cleavage by RNAse H. Formation of the AO–mRNA complex may also sterically hinder the access of ribosome to the mRNA, thus preventing translation.

AOs are produced by chemical synthesis and usually encompass 15 to 25 nucleotides, which ensures proper selectivity and strength of binding to mRNA. Natural oligonucleotides are susceptible to cleavage by exo- and endonucleases, which would terminate their action; therefore, synthetic AOs are modified so as to increase their stability.

There are already three generations of AOs obtained by successive chemical modifications. They are characterized by improved properties: stability and specificity. AOs are used to study the functions of genes and their products, and to some extent their use can be an alternative to transgenic animal experiments. They are also useful during drug target validation since by blocking the synthesis of a certain receptor protein, one can gain insight into the action of its potential low-molecular-mass antagonists. Thus, the possibility of eliminating particular proteins transiently allows us to mimic pharmacological intervention. The AOs themselves offer the great advantage over traditional drugs that they are very specific and prevent the production of proteins, while many conventional drugs inhibit the activity of proteins already present in an organism. The first antisense drug has already been marketed: the 21-nucleotides-long Vitravene, used for the treatment of retinitis caused by cytomegalovirus in AIDS patients. Other antisense drugs are currently undergoing clinical trials.

RNA Interference

RNA interference (RNAi) is a naturally occurring regulatory mechanism by which short (21 to 23 nucleotides long), endogenous, double-stranded RNA molecules, termed *short interfering RNA*, (siRNA) block the expression of genes bearing complementary sequences. This mechanism was first discovered in the nematode *C. elegans* but later it became clear that it is widespread and also occurs in mammals. SiRNA molecules are formed by processing of double-stranded RNA (dsRNA) or short hairpin RNA (shRNA) by an enzyme called Dicer. The newly formed siRNA molecules are incorporated into the RNA-induced silencing complex (RISC). It binds preferentially only the antisense siRNA strand, while the sense strand is degraded. The siRNA guides the RISC complex to the complementary mRNA sequence, which upon earmarking with siRNA, is cleaved, which prevents protein expression.

RNA interference appears to have great prospects of application in functional proteomics and potentially in therapy, although the latter is not yet available. Experimentally, gene expression is inhibited by a carefully designed exogenous synthetic siRNA duplex 21 nucleotides in length, whose sequence is fully homologous to the target mRNA. It is introduced into the cells by various transfection methods. Otherwise, the siRNA coding sequence can be engineered into a plasmid so that it can be expressed inside the cell, in which case its cellular effects last longer. The use of dsRNA molecules longer than 21 nucleotides would be likely to induce an immune response, naturally aimed at protecting an organism against viruses.

The main advantages of using siRNA for gene silencing are that it is highly efficient, the interfering molecules are readily available from many companies, and their efficiency of delivery into the cells is relatively high, although in mammalian cells the effects observed are transient, depending on the rate of cell division. Potential drawbacks to be considered include nonspecific interaction with mRNAs other than as planned; therefore, at the stage of experimental planning one should check whether homologous target sequences are present in other genes. Also, if too high a concentration of siRNA is used, the risk of nonspecific effects is even greater, as is the risk of evoking immunological response. Additionally, the process of transfection may affect the functioning of the cell, and transferring the siRNA agent into the correct target cell is not always trivial. Some of these disadvantages may be circumvented by chemical modification of exogenous siRNA in order to enhance its stability in the cell and to reduce off-target side effects by enhancing interaction with the mRNA sequence chosen.

It should be noted that siRNAs exploits the true physiological mechanism regulating cell functioning, which might be considered a disadvantage since this mechanism might be saturated, leading to unpredicted side effects. Knowledge regarding RNAi phenomena is growing rapidly; therefore, many improvements in our ability to exploit it can be expected in the future.

PROTEIN INTERACTION MAPS

High-throughput studies of protein interactions allow the creation of interaction maps (Figure 7) that facilitate understanding of crosstalk between different biochemical

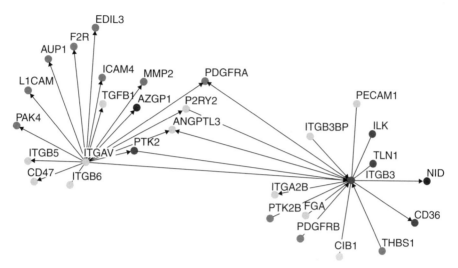

FIGURE 7   Fragment of a human interaction map obtained by proteomic examination of prostate cancer cell lines. (*See insert for color representation of figure.*)

pathways. They are a step on the way to defining the *interactome*, the set of all interactions that govern the functioning of a particular cell, tissue, or organism. A protein interaction map is not only a drawing that represents connections between particular proteins but represents real-life data obtained as a result of most complex studies. High-throughput functional proteomics experiments that lead to the construction of interaction maps may reveal novel roles of previously known proteins, propose new partners for disease-associated proteins, or integrate data pertaining to a particular biochemical pathway. All these advantages would not be possible without a careful bioinformatic and statistical analysis; therefore, to obtain significant results, concerted effort by interdisciplinary scientists is necessary.

ANIMAL STUDIES

In functional proteomics, many studies can be conducted using in vitro techniques that offer relative ease of data interpretation and can often be automated. Tissue cultures are a common tool, but they do not allow us to assess the role of particular proteins in the complex reactions of living organisms, such as anxiety or learning. Therefore, the use of laboratory animals, predominantly mice and rats, is in certain cases unavoidable. Some of the techniques described in this chapter may be conducted on live subjects (e.g., gene silencing), and indeed, for selected applications, functional proteomics has to resort to the use of animals. We share many physiological processes with rodents, and the results of such investigations may be used to design therapies for humans. All animal experiments must adhere strictly to ethical guidelines; they should be restricted to the lowest possible number of animals that are necessary to provide statistically significant data and must be conducted in a way that avoids animal suffering. To restrict the need for animal experiments, more and more methods for replacement of animal studies by tissue culture assays are developed and validated.

Studies of the functions of genes and their products can be conducted with transgenic animals: predominantly mice, less frequently, rats. This term relates to animals whose genome was modified by an additional DNA fragment, be it their own or that of another organism. Their genome may include an additional fragment that expresses a desired protein (knock-in), or an own gene modified such that it is not expressed at all (knockout). In the latter case, the gene may be modified in such a way that a truncated version of its product protein is expressed (e.g., lacking a selected domain). As a result, many types of tissues may be affected by the modification imposed.

Transgenic animals make it possible to assess the function of a modified gene and its products (proteins) in the functioning of the whole organism. Such studies are difficult and expensive. The efficiency of genome modification is low, and many transgenic embryos die before they are born. Although it can be informative regarding the significance of the new gene and its product for the development of the new organism, this also makes the procedure of obtaining adult transgenic animals most expensive. Since the function of the gene eliminated may be compensated by other physiological mechanisms, the results must be interpreted carefully and are usually accompanied by data obtained using other techniques.

PERSPECTIVES OF FUNCTIONAL PROTEOMICS

The aim of functional proteomics is to elucidate the functions of all proteins contained in living cells and to determine in what ways these functions depend on interactions with other proteins, posttranslational modifications, and outside conditions. In recent years we have witnessed enormous progress in this field, but there is still high demand for more and more robust and reliable methods of high-throughput proteomic research.

The techniques described here are commonly used, but the results obtained may be difficult to interpret and often require additional validation, which results from the great number of genes and the proteins they code for, the existence of alternative splicing variants, posttranslational modifications, and mutations. Additionally, the interpretation of huge amounts of data may pose special difficulties; therefore, further progress of functional proteomics is also dependent on the development of bioinformatic tools and standards to enable efficient and error-free analysis of results. Moreover, these bioinformatic tools should facilitate comparison and full exploitation of data obtained by different methods.

High-throughput functional proteomics have certainly taken an important place in life sciences, but it will probably long be paired with traditional biochemistry, which concentrates on the interactions of single proteins only. Still, functional proteomics facilitates gaining universal knowledge regarding various biochemical pathways and, together with structural proteomics, pushes forward the discovery of new drug candidates.

In recent years, several other disciplines have developed to complement unveiling the complex molecular backgrounds of life: metabolomics, peptidomics, lipidomics, transcriptomics, and so on. Together, and strongly supported by bioinformatics, they constitute the foundations of a novel concept of systems biology that tries to integrate various types of data in order to create a complete mathematical description of biological processes as a whole.

Acknowledgments

This work was supported by International Centre for Genetic Engineering and Biotechnology (ICGEB) grant CRP/POL05-02. Tomasz Dylag received a scholarship from the Foundation for Polish Science.

RECOMMENDED READING

Alterovitz G., Liu J., Chow J., Ramoni M.F. Automation, parallelism, and robotics for proteomics. *Proteomics* 6(2006) 4016–4122.

Bertone P., Snyder M. Advances in functional protein microarray technology. *FEBS J.* 272(2005) 5400–5411.

http://www.functionalgenomics.org.uk/sections/resources/protein_arrays.htm.

Mueller M., Martens L., Apweiler R. Annotating the human proteome: beyond establishing a parts list. *Biochim. Biophys. Acta* 1774(2007) 175–191.

Rana T.M. Illuminating the silence: understanding the structure and function of small RNAs. *Nat. Rev. Mol. Cell Biol.* 8(2007) 23–26.

Reid J.D., Parker C.E., Borchers C.H. Protein arrays for biomarker discovery. *Curr. Opin. Mol. Ther.* 9(2007) 216–221.

Reineke U., et al. Applications of peptide arrays prepared by the SPOT-technology, *Curr. Opin. Biotechnol.* 12(2001) 59–64.

Stelzl U., Worm U., Lalowski M., et al. A human protein–protein interaction network: a resource for annotating the proteome. *Cell* 122(2005) 957–968.

Uttamchandami M., Wang J., Yao S.Q. Protein and small molecule microarrays: powerful tools for high-throughput proteomics. *Mol. BioSyst.* 2(2006) 58–68.

Yu X., Xu D., Cheng Q. Label-free detection methods for protein microarrays. *Proteomics* 6(2006) 5493–5503.

10.4

CLINICAL PROTEOMICS

Anna Bodzon-Kulakowska

THE GOALS OF CLINICAL PROTEOMICS

The enormous advances that have been made in biology and medicine during the last few years allow for better understanding of the molecular basis of many processes that occur in living organisms. Pathological changes that lead to disease development are receiving special attention, as there are still many disorders, such as cancer, which lack effective treatment and remain a great challenge for clinicians. For the successful treatment of a disease, its early diagnosis is crucial, so there is an urgent need to identify and characterize specific biomarkers for a particular malfunction of an organism. The possibility of finding new targets for therapy and the chance to follow its progress are also very tempting. Genomics and proteomics, relatively new fields of molecular biology, seem to hold great promise for modern medicine.

Genomics offers two approaches for clinicians: genetic phenotyping and gene expression analysis. Genetic phenotyping is focused on hereditary predisposition to a disease, but mutation in a particular gene does not mean that a person will get ill. It signalizes only a risk for such predisposition. For example, mutation in BRCA1 (a tumor suppressor gene that plays a role in DNA repair) correlates with breast cancer, but 20% of persons who have experienced such a mutation remain cancer-free. Thus, after genome analysis, a person may be warned about the possibility of having a certain disease and may undertake appropriate precautions to monitor or even prevent development of a particular illness.

Proteomics: Introduction to Methods and Applications, Edited by Agnieszka Kraj and Jerzy Silberring
Copyright © 2008 John Wiley & Sons, Inc.

Gene expression analysis allows for determination of the level of gene transcript. In some cases, changes at this level may be associated with disease and could serve as a useful biomarker, but very often there is no such correlation. Gene expression may not necessarily be linked with the level of functional protein. What is more, its measure does not provide information about posttranslational modifications or C- or N-terminal truncation, which by changing the properties of protein, may be responsible for disease development.

Proteins possess a vast number of functions and participate in essentially all processes in living organisms; thus, it is not surprising that changes in their concentrations, or posttranslational modifications, can be the cause (or effect) of a disease. The majority of medications presently available are directed toward proteins. Proteomics can discover the dynamic changes in protein expression caused by a pathological state and is also able to capture changes in posttranslational modifications. Furthermore, the ability of proteomics platforms to observe a large number of proteins in a single experiment is of particular advantage. Such a molecular signature may allow for a more definite diagnosis or for a better assessment of the phase of a disease. Consequently, as proteomics can be adapted by clinicians to their studies, the term *clinical proteomics* was coined. There is a close feedback between basic proteomics, searching for potential markers, and clinical proteomics, dealing with protein profiles. But information gained in each of these approaches is constantly exchanged (Figure 1), thus leading to a better understanding of pathological processes and to closer collaboration among biochemists, mass spectrometrists, chemists, and physicians.

The practice of clinical proteomics should ensure earlier and/or more accurate diagnosis of a disease and should result in a better evaluation of its prognosis and prevention. It should also improve therapeutic strategies used to cure a particular illness. Clinical proteomics may also assure a less invasive or less risky diagnostic procedure. Currently, this field of science is focused primarily on diagnostics (i.e., finding new mechanisms and biomarkers for the disease), but it could also help in the identification of new therapeutic targets, drugs, and vaccines for successful disease prevention.

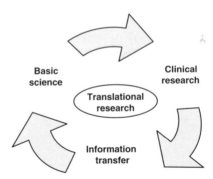

FIGURE 1   Relationship between basic and clinical proteomics.

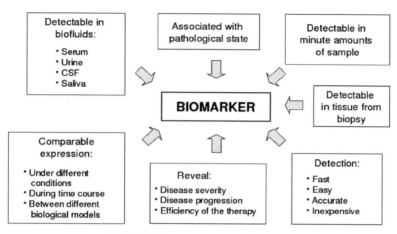

FIGURE 2   Features of a biomarker.

A FEW WORDS ABOUT BIOMARKERS

Clinical proteomics is often associated with the search for protein biomarkers of a disease; thus, a question arises regarding their correct definition. Several features of a potential biomarker (Figure 2) are as follows:

- A biomarker is a protein whose expression is associated specifically with a particular pathological state (e.g., level affected, modifications). The presence of this protein can be used to indicate the disease or to distinguish among its variants.
- A biomarker should be able to reveal the severity of a disease, its progression, and the efficiency of the therapy.
- It should be possible to detect this protein in an easily accessible biological fluid such as serum, urine, saliva, or in the case of cancer, in the tissue from biopsy.
- The amount of sample sufficient for analysis should be minute, and detection of a particular biomarker should be easy, fast, accurate, easy to automate, and inexpensive.
- The expression of such a characteristic protein should be comparable under a variety of conditions, during the time course, and between biological models: for example in vitro in cell culture and in vivo.

Biomarkers can be divided into three groups (especially in the case of cancer): diagnostic, prognostic, and predictive. *Diagnostic biomarkers* assist in the early detection of disease. They may also facilitate histopathological tumor classification. Their role is very important, as accurate tumor diagnosis is necessary for the choice of optimal therapy. *Prognostic biomarkers* provide information about the malignant potential of tumors, which is very valuable for further treatment. For example, if a certain biomarker indicates the possibility of recurrence, additional therapy could be implemented, whereas a patient without such a prognosis could be excluded from treatment. *Predictive biomarkers* are able to distinguish between different types of certain cancers and thus allow for proper selection of a successful therapy. For example, breast cancer that exhibits estrogen receptor–positive tumors will need different therapy (with antiestrogen compounds) than breast cancer that is estrogen receptor negative (it should be treated with chemotherapy).

Sometimes, a single biomarker is sufficient to describe the state of an organism. For example, human β-chorionic gonadotropin is used as a single diagnostic biomarker of pregnancy and is characterized by high specificity and sensitivity. In other situations this seems to be insufficient because the biology of the pathological state is very complex. In addition, due to the genetic heterogeneity among populations, a single biomarker which works for one group of people may fail for another. Moreover, the possibility of combining several biomarkers may result in a more accurate diagnosis.

> The main idea behind clinical proteomics is *not* identification of all components but rather, finding specific markers that can be used in clinical diagnosis beyond all existing tests in clinical chemistry. The key difference is that we search for the unknown: new mechanisms and pathways involved in a particular disease, and we carry this out on a mass scale.

Biological material suitable for clinical analysis, usually comprised of bodily fluids such as plasma/serum, synovial fluid, saliva, or cerebrospinal fluid, is very complex to use for detailed and fast analysis, including full identification of all components. Clinical proteomics strategy compares on a basic level, protein and peptide profiles without full identification of particular compounds. The rationale is that in a clinical environment, samples need to be analyzed with appropriate speed (not exceeding few hours) and high throughput (Figure 3). This resembles typical clinical chemistry

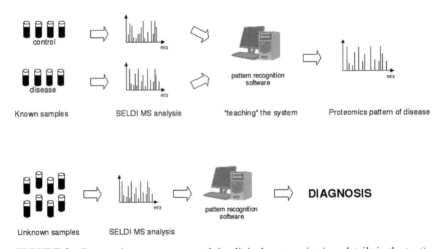

FIGURE 3 Proteomics pattern approach in clinical proteomics (see details in the text).

analysis with one difference: The entire patterns are compared instead of a single component such as the glucose level for early detection of diabetes.

Protein profiles form control samples, and those obtained from patients are analyzed with special pattern recognition software to create the proteomics pattern of a disease (see the detailed description below and Chapter 11). Such a molecular signature may be characteristic for a pathological state and be sufficient for its initial diagnosis. However, further characterization of protein biomarkers present in such a pattern may have a lot of benefits. First, this identification may improve our knowledge of the mechanism(s) underlying a certain disease and its pathophysiology. It could also facilitate new methods for fast diagnosis (e.g., ELISA techniques), as proteomics techniques are time consuming, quite expensive, and demand expensive laboratory equipment.

Apart from the protein profile approach, "classical" comparative proteomics is used to find the differences in protein expression between different states of an organism and could also be used to find biomarkers of the disease (Figure 4). It uses classical proteomics methods such as two-dimensional gel electrophoresis and multidimensional LC-MS to find those proteins and is focused on their identification. Such identified biomarkers may be then employed for clinical use and diagnosis by, for example, ELISA.

The work flow for implementation of the appropriate biomarkers of particular disease in clinical use should be as follows: discovery of potential biomarkers, their validation due to their ability to make useful predictions in patient populations, and implementation (i.e., development of clinical assay) (Figure 5).

The discovery of biomarkers may be facilitated by a careful choice of material where the pathological changes may occur. As we know that prefractionation of

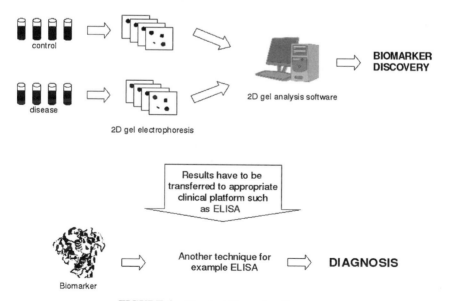

FIGURE 4   Classical biomarker discovery.

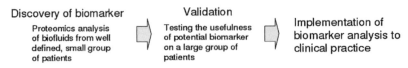

Discovery of biomarker

Proteomics analysis
of biofluids from well
defined, small group
of patients

Validation

Testing the usefulness
of potential biomarker
on a large group of
patients

Implementation of
biomarker analysis to
clinical practice

FIGURE 5   Work flow for implementation of a biomarker in a clinical trial.

biological samples allows for observation of low-abundant proteins in proteomic studies, such an approach may be very fruitful. In this step, identification and quantification of particular protein biomarker should be achieved.

After that, the ability of a presumptive biomarker or biomarkers to distinguish between a pathological state and a normal condition has to be measured in validation studies. This ability is characterized by sensitivity and specificity. Sensitivity means the ability to detect true positives and is expressed as a percentage expressing the number of positives detected relative to the total number of patients with a disease. Specificity measures the ability to eliminate false positives (a percentage of positive results that were actually true positives). The group of patients at this stage of study should be much larger than the one used at the beginning of the study and independent of the first group. The assay that is used at this stage should be the same as the one proposed for implementation.

TECHNIQUES AND METHODS USED IN CLINICAL PROTEOMICS

Proteomics is able to ensure less invasive and/or risk-associated procedures for diagnosis or therapy evaluation. But to be useful in clinical practice the proteomics platform for defined procedure should be chosen carefully.

Two-dimensional gel electrophoresis and multidimensional LC-MS are extremely useful in the discovery of potential biomarkers. In case of two-dimensional electrophoresis, protein samples obtained from the patients with a certain disease and those from control persons have to be separated on the gels and stained with Coomassie Brilliant Blue, silver, or fluorescent dyes. The staining intensity of spots on the gel could be measured, and thus differences in expression between two samples may be found. Spots corresponding to proteins whose expression is different due to their pathological state are then excised, the proteins are digested (usually, by trypsin), the peptides obtained are extracted from the gel, and the sample so prepared is analyzed by mass spectrometry. A protein may be identified on the basis of its mass fingerprint MS spectrum in protein databases, and thus a potential biomarker of that pathological state may be identified (for details, see Chapters 7.1 to 7.3).

Multidimensional LC-MS may also serve for the separation of proteins and the identification of possible differences. A sample representing pathological state may be labeled with an isotope tag and its content then separated simultaneously with nonlabeled control on a liquid chromatography system (LC). A difference in intensities between peaks from labeled and nonlabeled peptides may indicate a potential biomarker (see Chapter 7.6).

Apart from their benefits, those two techniques are extremely time consuming and therefore may not find direct application in clinical practice, and the results have to be transferred to an appropriate clinical platform such as ELISA.

Techniques based on MALDI or SELDI seem to be more proper because of their ability to perform fast analysis. Apart from protein identification, MALDI may be used for entire protein pattern measurements. The procedure for MALDI analysis is very fast; it demands mixing the special matrix with the protein mixture and performing several laser shots to obtained protein–matrix crystals to acquire a spectrum. Because of the sensitivity of this method, very low concentrations of proteins may be measured. This is important, as biological biomarkers are usually in very low concentration.

SELDI-TOF-MS or ClinProt technologies are the most useful in clinical applications. For SELDI analysis the sample (e.g., a biofluid) is placed on the ProteinChip array surface. This modified surface has certain physicochemical characteristics (hydrophobic, hydrophilic, cationic, anionic, or metal ion) to adsorb proteins selectively, which results in a reduction in sample complexity. All contaminants and proteins that were not captured are then removed and a protein sample so prepared is mixed with matrix and analyzes, as in the case of MALDI. SELDI approach reduces the complexity of the sample which results in the probability of detecting the markers that are present at low concentrations.

A final method that could be used in clinical proteomics is one based on magnetic beads. Those beads are used for selective capture of proteins and peptides within a sample. They may have anionic, cationic, hydrophobic, normal phase, or metal coverage. Magnetic beads are added to the sample solution and target proteins are bound to them based on their absorption to the ligand, which results in enrichment of a specific subset of proteins. Beads with certain proteins are then separated from the sample using a magnetic bead separator. The proteins may then be eluted, mixed with matrix, and measured by MALDI-MS, which could be used here for protein identification. The protein and peptide interaction is also measured here.

The advantage of MALDI, SELDI, or ClinProt is the short analysis time required by these methods, their sensitivity (picomole to femtomole range), and the small amounts of sample necessary for analysis. The proteomics patterns obtained here may serve in disease diagnosis. One problem with such an analysis is that they are based on a comparison of profiles and not on detailed identification of protein sequences. Therefore, it is difficult to prove that a certain peak on the spectrum corresponds to a particular protein. The data obtained need to be verified by more detailed studies, which are available using two-dimensional gel electrophoresis or multidimensional LC-MS.

SAMPLE SOURCE AND PROTEOMIC PATTERN APPROACH IN CLINICAL PROTEOMICS

Bodily fluids such as plasma, urine, and cerebrospinal fluid are very attractive as a source for clinical analysis. They all carry a lot of vital information about the condition of an organism and can be collected relatively easily.

One way in which proteomics could be used in clinical diagnosis is by creating a proteomic pattern of a disease from biological fluids (see Figure 3). Pattern recognition software based on artificial intelligence algorithms and chemometrics (see also Chapter 11) is "taught", using profiles from healthy and ill persons, what the protein profile of healthy person looks like and the difference between this pattern and the one obtained from a patient with a pathological state. Thus, a proteomic pattern (or fingerprint) of the disease—the peak combination that is characteristic for a pathological state—is created.

Such a proteomics pattern comprises many individual proteins. Each of these proteins may independently comprise a marker of the disease; together, they create the pattern characteristic for, the pathological state. Moreover, these proteins may not necessarily be connected with a pathological state (e.g., in the case of cancer, they may not be tumor-derived products). They may reflect metabolic changes occurring due to disease development.

Below, short descriptions of biological fluids generally used as a source of protein biomarker mining are presented, together with examples of clinical studies. An approach to solid tissue analysis is also noted.

Plasma and Serum

Plasma is a liquid portion of blood, without blood cells; *serum*, additionally, has no platelets or clotting factors. Serum has a very high protein concentration (Figure 6), but 99% of its protein content belongs to the 22 most abundant molecules (e.g., albumin, haptoglobins, immunoglobins, transferrins). Only the remaining 1% consists of lower-abundance proteins (regulatory proteins), which are the main focus of clinical proteomics. Some approaches try to eliminate those abundant proteins by immunodepletion, for example, but care must be taken, as it was shown that some clinically relevant biomarkers may be bound to highly abundant circulatory proteins such as albumin. Taking everything into account, plasma is one of the most difficult proteomics challenges to characterize because of a large proportion of highly abundant proteins and a very wide dynamic range of protein concentration equal to 10^{10}.

During circulation, blood may collect certain proteins from the sites in an organism affected by a disease. Thus, apart from the classical plasma protein (e.g., protein

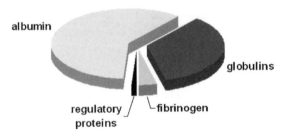

FIGURE 6   Protein content of plasma: albumin (ca. 60%), globulins (ca. 35%), fibrinogen (ca. 4%), and regulatory proteins less than 1%.

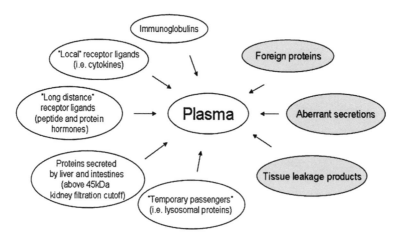

FIGURE 7   Sources of plasma proteins. Those indicated in gray are pathological.

secreted by liver and intestines, immunoglobulins, peptide and protein hormones, cytokines), it could contain potential biomarkers of that disease (Figure 7). Among them could be tissue leakage products—proteins that are normally present in the cell, but due to cell death or damage, are released. Another group may involve aberrant secretion, which usually takes place in cancer cells or damaged tissue. Proteins secreted in such a way are not present in a healthy circulation system. Usually, they are normal nonplasma proteins. The last group consists of "foreign" proteins, expressed or released by infectious organisms.

Adam et al. (2002) used a protein pattern approach to make possible early detection of prostate cancer (PCA) from serum analysis. Nowadays, a serum PSA (prostate-specific antigen) test is used for detection of this cancer, but the positive result still has to be confirmed by biopsy. Despite its high sensitivity (more than 90%), this test lacks specificity (only 25%), which sometimes means unnecessary biopsies; thus, the search for a new biomarker of this disease seems to be indispensable.

Serum samples from 77 patients with benign prostate hyperplasia, 167 with prostate cancer, and 82 from age-matched, unaffected, healthy men were subjected to SELDI-MS analysis. The spectra obtained were used to train an algorithm to differentiate between normal and pathological states. At the end the system was able to distinguish between a normal and a pathological state with 83% sensitivity and 97% specificity.

Urine

Due to its proximity, urine represents a source of proteins that could act as potential biomarkers of urinary tract diseases. In addition, its proteome (and metabolome) could be affected by drug treatment; thus, its analysis may provide further information related to drug metabolism. This biofluid is attractive from a clinical point of view because it may be collected easily and noninvasively.

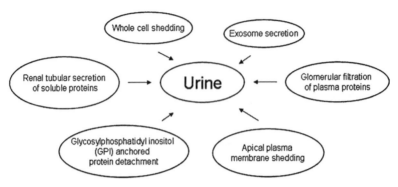

FIGURE 8 Sources of urine proteins.

Urine contains proteins that may originate from glomerular filtration of plasma proteins, renal tubular secretion of soluble proteins, whole-cell shedding, apical plasma membrane shedding (nonspecific or apoptotic processes), glycosylphosphatidyl inositol (GPI)–anchored protein detachment, or exosome secretion (Figure 8). Changes in urine concentration of those proteins may result from a change in their content in blood plasma, or from functional changes in the glomerular filter or in the proximal tubule scavenging system, and thus indicate pathological processes in these systems. As in plasma analysis, highly abundant proteins [filtered from plasma or produced by renal tubular cells (e.g., the Tamm–Horsfall protein, a glycoprotein, which is the major fraction of the uromucoprotein)] make analysis of low-abundant species difficult.

Urine as a clinical sample creates another problem; it should be collected for 24 hours for quantitative measurement of the rate of excretion of putative biomarkers. Measurement of biomarker concentration in a single spot of urine is insufficient because normal physiological changes in water excretion can dilute or concentrate urinary proteins. As 24 hours of urine collection may be difficult and not convenient for patients, an alternative is normalization of the biomarker concentration. This could be done by normalization of biomarker concentration to the concentration of creatine, insulin, or other filterable but not reabsorbable markers that are excreted in the urine at relatively constant rates. One should also keep in mind the possibility of bacterial contamination in the first morning urine due to long-term resistance in the bladder.

In one study, urine was used to detect potential biomarkers of transitional cell bladder carcinoma (TCC). At present, this cancer may be detected only by invasive methods such as cytoscopic examination (an internal examination of the bladder) and biopsy. Vlahou et al. (2001) detected five prominent protein peaks and seven protein clusters on the spectra of urine from patients with TCC which could serve as potential indicators of this cancer.

Cerebrospinal Fluid

Cerebrospinal fluid (CSF) protects the central nervous system from changes in blood pressure and trauma. Neural cells shed proteins and peptides into extracellular fluid. CSF carries neurosecreted, biosynthesized, and metabolized cellular products, and thus acts as a protein repository in the brain. Because of the presence of the blood–brain barrier, there is no direct exchange of molecules between CSF and the peripheral circulatory system. As this fluid is in direct contact with the central nervous system (CNS), it may serve as a source of potential biomarkers of neurological diseases.

One challenge in the analysis of CSF proteome is the low abundance of proteins, peptides, and proteolytic fragments comparised with serum. Among the most abundant proteins in CSF are albumin, immunoglobulins (Igs), and transferrin. Two-dimensional gel electrophoresis revealed more than 500 protein spots during analysis of this biofluid. CSF protein concentration increases with most brain diseases, choosing this fluid as a possible source of information about pathological processes in the CNS.

As an example, some trials were performed to distinguish Alzheimer disease from other dementia cases, and to differentiate changes associated with age from those caused by pathological changes. Thus, there is an urgent need for a valid test for this disease. As there is no single biomarker for such neurodegenerative diseases, attempts were made to find a specific pattern of changes in a set of proteins from CSF. Castano et al. (2006) used two-dimensional gel electrophoresis of CSF from patients with Alzheimer disease and controls, followed by MALDI identification. They found five proteins with potential roles in amyloid-beta metabolism and vascular and brain physiology and two pigment epithelium-derived factors that discriminate between those two groups.

These results indicate that pattern analysis seems to be more promising than the search for a single biomarker. In clinical application serum seems to be the most promising source of information. Withdrawal of cerebrospinal fluid is always associated with a sometimes painful and risky invasive clinical procedure, and urine is difficult to prepare in a short time.

Tissue Analysis

Clinical proteomics is not focused solely on bodily fluids. Recent findings suggest that it may support histopathological examination in discriminating between normal and malignant tissues, which is a crucial step in cancer diagnosis. A comparative proteomics study of material from biopsy may indicate proteins which, apart from being potential biomarkers of a malignant state, could become a target for therapeutic agents in the future.

Studies by Chen et al. (2003) were focused on lung adenocarcinoma. Nearly 35% of patients with stage I adenocarcinoma will not survive five years; for the remainder, the prognosis is more optimistic. By using two-dimensional gel analysis followed by MS, 682 individual protein spots were analyzed in 90 lung adenocarcinomas. During this study, 20 potential biomarkers associated with patients' survival were found. Those biomarkers may lead to the selection of high-risk patients and may allow for application of additional therapy, apart from surgical resection, to increase survival. It is interesting that one of the proteins associated with survival prediction, phosphoglycerate kinase 1, was found in serum when using ELISA, which indicates that proteomics study of cancer tissue may be helpful in finding corresponding

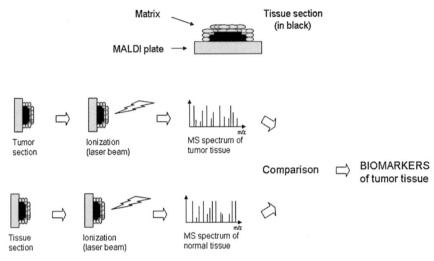

FIGURE 9 Discovery of tumor tissue biomarkers using imaging mass spectrometry.

biomarkers in biofluids such as serum, where they are easily measurable. It should be noted, however, that a particular molecule might serve as a predictive marker in certain cells but not elsewhere.

Apart from the classical proteomic analysis, a new technique, imaging mass spectrometry, has been introduced for tissue analysis (Figure 9). Here we can observe many molecular species simultaneously. More details on this technique and its possibilities can be found in Chapter 7.3.

FUTURE PERSPECTIVES

In the case of cancer, where clinical proteomics is used most intensely, there is a dictum which says that "no two cancer patients are alike," because patients who have the same malignancies, disease sites, and precipitating factors can and do respond differently to treatment with the same agents. Application of proteomics in clinical use may, in the future, result in a "patient-tailored" therapy or "personalized molecular medicine." In this approach, the disease could be detected early and classified by proteomic profiling of biological fluid. Then on the basis of such profiling, appropriate therapeutic agents may be used. Advances in therapy may be monitored and fine-tuned by proteomic pathway profiling. We may anticipate the practical implementation of imaging mass spectrometry to evaluate margins during surgical resection of the tumor, as its classical histopathological analysis is very difficult. The sensitivity of mass spectrometry may assure better removal of tumor tissue than is now achieved.

Although there are still a lot of challenges for proteomics to traverse before it will be used routinely by clinicians, it will definitely change the practice of medicine. We will then be able to provide individualized and "best ever" therapy for patients.

Acknowledgments

This work was supported by International Centre for Genetic Engineering and Biotechnology (ICGEB) grant CRP/POL05-02.

RECOMMENDED READING

Adam B., Qu Y., Davis J.W., Ward M.D., Clements M.A. Serum protein fingerprinting coupled with a pattern-matching algorithm distinguishes prostate cancer from benign prostate hyperplasia and healthy men. *Cancer Res.* 62(2002) 3609–3614.

Anderson N.L., Anderson N.G. The human plasma proteome: history, character, and diagnostic prospects. *Mol. Cell Proteom.* 1(2002) 845–867.

Azad N.S., Rasool N., Annunziata C.M., Minasian L., Whiteley G. Proteomics in clinical trials and practice: present uses and future promise. *Mol. Cell Proteom.* 5(2006) 1819–1829.

Castano E.M., Roher A.E., Esh C.L., Kokjohn T.A., Beach T. Comparative proteomics of cerebrospinal fluid in neuropathologically-confirmed Alzheimer's disease and non-demented elderly subjects. *Neurol. Res.* 28(2006) 155–163.

Chaurand P., Sanders M.E., Jensen R.A., Caprioli R.M. Proteomics in diagnostic pathology: profiling and imaging proteins directly in tissue sections. *Am. J. Pathol.* 165(2004) 1057–1068.

Chen G., Gharib T.G., Wang H., Huang C., Kuick R. Protein profiles associated with survival in lung adenocarcinoma. *Proc. Natl. Acad. Sci. U. S. A.* 100(2003) 13537–13542.

Mischak H., Apweiler R., Banks R.E., *et al.* Clinical proteomics: a need to define the field and to begin to set adequate standards. *Proteom. Clin. Appl.* 1(2007) 148–156.

Romeo M.J., Espina V., Lowenthal M., Espina B.H., Petricoin E.P., Liotta L.A. CSF: a protein repository for potential biomarker identification. *Expert Rev. Proteom.* (2005) 57–70.

Seibert V., Ebert M.P.A., Buschmann T. Advances in clinical cancer proteomics: SELDI-TOF-mass spectrometry and biomarker discovery. *Brief Funct. Genom. Proteom.* 4(2005) 16–26.

Varambally S., Yu J., Laxman B., Rhodes D.R., Mehra R. Integrative genomic and proteomic analysis of prostate cancer reveals signatures of metastatic progression. *Cancer Cell* 8(2005) 393–406.

Veenstra T.D., Conrads T.P., Hood B.L., Avellino A.M., Ellenbogen R.G. Biomarkers: mining the biofluid proteome. *Mol. Cell Proteom.* 4(2005) 409–418.

Vlahou A., Schellhammer P.F., Mendrinos S., *et al.* Development of a novel proteomic approach for the detection of transitional cell carcinoma of the bladder in urine. *Am. J. Pathol.* 158(2001) 1491–1502.

Yanagisawa K., Shyr Y., Xu B.J., *et al.* Proteomic patterns of tumour subsets in non-small-cell lung cancer. *Lancet* 362(2003) 433–439.

Zhang X., Leung S., Morris C.R., Shigenaga M.K. Evaluation of a novel, integrated approach using functionalized magnetic beads, bench-top MALDI-TOF-MS with prestructured sample supports, and pattern recognition software for profiling potential biomarkers in human plasma. *J. Biomol. Technol.* 15(2004) 167–175.

11

BIOINFORMATICS

Pawel Ciborowski and Hesham Ali

Technological advancements in the 1980s and 1990s opened new possibilities in the study of changes in biological systems at the global level. Development in areas of mass spectrometry of proteins and peptides as well as the advancement of two-dimensional electrophoresis into a more reproducible and user-friendly method laid the foundation for the new field of proteomics. In consequence, combined global proteomic and genomic approaches, along with steadily increasing computer-processing capabilities, made it possible to accumulate unprecedented amounts of new and original data within a short period of time. This created a demand for better and faster tools to transform data to information and subsequently to knowledge that sets new challenges for bioinformatics. Tremendous advancement in computational sciences has transformed many scientific disciplines, including life sciences, not just by providing more data and speeding up steps traditionally done without automation, but by providing novel and powerful perspectives which often lead to unforeseen insights.

With a strong foundation in computational biology and computational chemistry, bioinformatics is emerging fast as one of the most exciting scientific disciplines in the twenty-first century. This undoubtedly interdisciplinary field of study deals with the development and use of mathematical and computational methods to assist in modeling and solving problems in the life sciences. Employing algorithmic approaches is expected to be essential to development and advancement in many bioscience fields. This, in part, is due to the recent explosion of biological data, which requires a coordinated increase in the scale and sophistication of the automated systems and tools that enable researchers to take advantage of its availability. Additionally, there are a large number of research projects and applications that demand automated computational support.

Proteomics: Introduction to Methods and Applications, Edited by Agnieszka Kraj and Jerzy Silberring
Copyright © 2008 John Wiley & Sons, Inc.

The growing need for bioinformatics reflects the radical changes in the biological sciences over the last decade. As more important biological elements are studied and their roles in complex biological systems are discovered, it is apparent that integrating computational research and experimental work will be crucial in exploring and understanding these discoveries. Also, as more biological data become available to researchers, efficient data-mining algorithms and data curation techniques will be essential in order to explore the data available and extract valuable information. Further progress in biological disciplines such as structural biology and functional genomics will not be possible without computational methods and bioinformatics tools.

Bioinformatics is a relatively young scientific discipline and one can relate its birth to creation of the ARPANET linking computers at Stanford, the University of California at Santa Barbara, the University of Utah, and UCLA in 1969. This was followed by such events as the development of the Needleman–Wunsch algorithm for sequence comparison in 1970 and the invention of e-mail in 1971 (http://openmap.bbn.com/~tomlinso/ray/firstemailframe.html). Further development of bioinformatics was precipitous. How should we define this field?

There are many proposed definitions of bioinformatics and continuously ongoing debates about how to describe this emerging scientific field in the most complete and comprehensive way. One definition is that *bioinformatics* (computational biology) is a highly interdisciplinary scientific discipline which uses tools, techniques, and concepts from applied mathematics, informatics, statistics, chemistry, biochemistry, physics, linguistics, and computer science to solve biological problems. The authors are aware that this definition does not exhaust all aspects; however, it gives a sense of the interdisciplinary nature of bioinformatics. Another question is: What do we expect from bioinformatics? In our view, bioinformatics develops methods of data storage, retrieval, and analysis as well as sequence alignment, protein structure prediction, prediction of gene expression, protein–protein interactions, and modeling of protein networks. In an effort to better define the field of bioinformatics and its subdisciplines, a distinction has been drawn between bioinformatics and biomedical informatics, also called *medical informatics*, a subdiscipline narrowed to understanding, managing, and communicating medical information using the same tools as other subdisciplines of bioinformatics, such as algorithms, statistics, and network modeling.

One major aim is to improve communication between those who deposit data and those who want to utilize information provided by others. Therefore, biomedical informaticians are more focused on data itself and biological (medical) utility, while bioinformaticists are focused on the theory supporting data manipulation.

With the advent of faster and more reliable technology for sequencing nucleic acids and proteins, very large centralized biological databases have become available. The massive size of the biological databases available and its rapid growth have influenced the types of research currently being conducted, and researchers are focusing more than ever on maximizing the use of the biological databases available. Because biology has a long tradition of comparative analysis leading to knowledge and discovery, it would be a great advantage for bioinformatics researchers to utilize the information stored in the available databases to extract new biological information as well as to understand various biological phenomena.

Although bioinformatics may benefit from employing existing computational techniques to address various problems in the biosciences, for bioinformatics to emerge as an independent and useful field of study, novel research needs to be developed. Such research has to be more than direct and simple in the use of well-known pattern recognition and data-mining algorithms to problems in biosciences as conducted primarily by computational scientists. By the same token, future bioinformatics research also has to be more than the simple use of general tools to explore similarities among data in the biosciences domain as performed daily by bioscientists. In other words, new advanced research in areas such as artificial intelligence, data mining, and information retrieval in the specific domain of bioinformatics will be critical to the development of bioinformatics research.

In summary, the new discipline of bioinformatics provides an excellent opportunity for researchers to cross the bridges connecting biosciences and computational sciences. Only genuine collaboration among researchers will lead to effective utilization of the advancement of technologies and allow the various fields of biosciences the desired acceleration to achieve many of their objectives.

SYSTEMS BIOLOGY AND BIOINFORMATICS

One inherent property of new and breakthrough technologies is the constant demand for further and faster expansion. It is fascinating how fast the field of proteomics is evolving in this respect. Proteomics emerged as a scientific field in the mid-1990s along with technological advances in the analysis of proteins and peptides. Technologies such as two-dimensional polyacrylamide gel electrophoresis (SDS-PAGE), developed originally in the late 1970s (Barritault et al., 1976), and advanced rapidly to the next level when immobilized pH-gradient (IPG strips) gels were introduced (Gelfi and Righetti, 1983), facilitating analysis of gene products containing more than 2000 proteins in a mixture (Gianazza, 1995) with much higher reproducibility. Isotope tags for relative and absolute quantification (ITRAQ), isotope-coded affinity tags (ICAT), and stable isotope labeling with amino acids in cell culture (SILAC) gave an even further technological boost and opened new avenues of studying functions of organisms at a global level. Increasing throughput of profiling experiments along with growing computer-processing capabilities caused still more demand for additional global studies. This advancement quickly led to development in the discipline of systems biology. There are many definitions of this field, and we adopted one proposed by Wikipedia, which states that "systems biology is a field of study in the biosciences that focuses on the systematic study of complex interactions in biological systems" (http://en.wikipedia.org/wiki/Systems_biology). Although the history of systems biology is only about four decades old, it has thrived in recent years. It is unquestionable that this discipline will deliver knowledge that will advance tremendously our understanding of life at the molecular level. However, building knowledge based on accumulated information seems to be the most difficult task. Bioinformatics using various tools is quickly advancing our ability to transform data into information and to compare pieces of information of the same type, such as two-dimensional gels and mass spectra. What remains a bottleneck is to integrate information from various platforms and *-omics* approaches (i.e., genomics, proteomics, cellomics, metabolomics, phylomics, transcriptomics, etc.).

Two general approaches have been proposed to study systems biology. The *top-down approach* deals with system modeling leading to hypothesis, prediction, and simulation of the behavior of an organism under various physiological conditions. In the *bottom-up approach* we study individual pieces on a discovery basis and combine all the information to understand how the system in question works. Figure 1 presents this concept. There is no question that both approaches have the same goal, which is

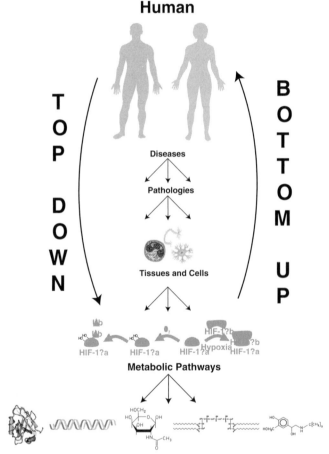

FIGURE 1 Top-down and bottom-up approaches to systems biology. There are components that contribute equally to the study of biology systems: theoretical, conceptual, and experimental. All of these approaches depend more and more heavily on implementation of effective bioinformatics tools. Hypothesis- and discovery-driven projects participate equally in studying systems.

to describe the function of a biological system as one entity. Handling mixtures of peptides generated from a complex protein mixture (the bottom-up approach) is very different from sample processing at the protein level prior to analysis (the top-down approach). On the other hand, the mass spectrometry of protein peptides, as well as other technologies used, is similar.

Nevertheless, the top-down strategy requires high-end instrumentation with high mass accuracy and resolution, using isolation, analysis, and characterization of an intact protein as an experimental approach to reveal its function. Fourier-transformed/ion cyclotron resonance mass spectrometry (FT-ICR) (Marshall et al., 1998) facilitates such an approach in protein identification as a result of random fragmentation of intact molecules. The bottom-up approach, which relies more on front-end fractionation, may also use mass spectrometers which have lower mass accuracy and resolution.

To some extent, there is a difference in bioinformatics methodology associated with each of these two approaches. For example, protein identification using high-mass-accuracy FT-ICR will use a specific algorithm, searching the same databases as were searched by other algorithms. The bottom-up approach will also use the same databases, however, it requires up-front fragmentation of the protein in question using various proteolytic enzymes with known specificity (Chalmers et al., 2005; Millea et al., 2006). Eventually, experimental design and questions asked by researchers will decide which approach is more appropriate for a project and whether more information will be obtained from the bottom-up approach (protein identification at a cost of protein characterization) or the top-down approach (distinguishing differences between two similar proteins or isoforms). It would be ideal to obtain all information from one type of analysis. This technological challenge is also a challenge for bioinformatics: to facilitate integration of information and the building of knowledge.

HIGH-THROUGHPUT PROTEOMICS

Proteomics is envisioned similar to genomics: that is, a high-throughput experimental approach. As this is certainly true for genomics, proteomics throughput has a different dimension. Available definitions of the term *high throughput* can be interpreted in somewhat different ways. One definition (http://www.doylefoundation.org/icsu/glossary.htm) highlights "screening [that] makes use of techniques that allow for a fast and simple test on the presence or absence of a desirable structure" while an other definition (http://www.med.umich.edu/genetics/glossary/) highlights "the use of automated equipment and powerful searching software to tackle large and complex problems in genome and proteome studies." Thus, what is critical for the high-throughput approach: speed and ease or automation? Probably both. A critical factor in high-throughput proteomics (HTP) will be automated validation (Hogan et al., 2006) of data because both manual interpretation of large data sets and formats on an information output are impossible. The second factor is not unique for HTP because regardless of the size of proteomic experiments, interpretation of data and building knowledge will face similar challenges and questions.

Sample preparation ⟶ Measurements ⟶ Data processing

FIGURE 2   Schematic flow of a typical proteomic experiment, either low or high throughput. In low-throughput experiments an investigator can spend more time analyzing individual samples, while in HTP, investigators have no choice and must relay on fully automated protocols.

How can available technology platforms meet the criteria of speed, simplicity, and automation of high-throughput proteomic experiments in a simplified scheme of experimental work flow as shown in Figure 2? How many proteins (peptides) are not identified in any given high-throughput experiment, or how much information do we miss? If more than one algorithm is used to search databases, which criteria should be used? What is a definition of a reference sample itself, and how should it be used for standardization of data sets in HTP? Where is the bottleneck?

Reproducibility of sample preparation determines the quality of mass spectrometry measurements and later, the quality of data interpretation. Therefore, sample preparation, including sample acquisition, has to be well controlled. A sample of cerebrospinal fluid (CSF) is a very good example. If hemoglobin is detected in a CSF sample, even without the visual presence of blood, it means that a blood vessel was punctured during the spinal tap, and such a sample is contaminated with plasma proteins. In consequence, this may have a great impact on the ratio, in particular, low-abundant proteins in the system studied, in particular in experiments utilizing the ITRAQ and ICAT methods of protein labeling (tagging). This example shows that reproducibility (sample uniformity) starts even before a sample reaches the laboratory. Despite the growing number of proteomic studies of CSF, there are no uniform rules about how to deal with such issues; thus, comparison of data sets among laboratories and centers is very difficult.

Increasing high throughput in the crystallization of proteins for x-ray studies is another example of a challenge in global studies (McMullan et al., 2005). The Joint Center for Structural Genomics (http://www.jcsg.org/) is a multi-institutional consortium supported by the National Institutes of Health (www.nih.gov) and its goal is "to establish a robust and scalable protein structure determination pipeline that will form the foundation for a large-scale effective production center for structural genomics." The accomplishments of this center are certainly impressive, yet they show how much work remains to be done.

It is difficult to imagine global studies such as an investigation of systems biology without multiplexing. Current and future progress cannot advance fast enough without multiplexing in data acquisition and more efficient data analysis. Measurement of 10 parameters will generate a question of how 20, 50, or 100 parameters will change in a system in question. Thus, multiplexing appears as an option of choice. Generally, multiplexing can be understood or defined as multiple assays (measurements) performed at the same time using one analytical method, which is an approach of choice to increase throughput in analysis of complex mixtures. One example of a multiplexing experiment widely used in proteomics is two-dimensional electrophoresis with DIGE (difference gel electrophoresis), which utilizes labeling control and experimental proteins with various cyanine dyes. Although multiplexing is a key approach with an almost endless list of applications, the design of a multiplex assay is a challenging task. It requires compromise among objectives of such experiments, which might be competitive in nature. Another challenge is analysis and alignment of data (results), which always requires high computer-processing capabilities and custom-made software. The advantages and limitations of multiplex experiments apply to proteomics and genomics studies in the same way. The difference will only be applied technology. Table 1 presents some characteristics of the most commonly used profiling techniques.

Although recent years reflect the fast development of hyphenated method (combination of two analytical methods, e.g., liquid chromatography and mass spectrometry), the comparison presented in Table 1 shows limitations in proteomic throughput. Still, multiple instruments are often the only way to increase the number of samples analyzed. From this perspective, protein arrays appear as the most promising platform.

Has HTP matured technologically and bioinformatically enough to generate new and biologically relevant data? There is no one easy answer to this question, and

TABLE 1   Strengths and Weaknesses of the Most Commonly Used Profiling Methods

Method	Sample Preparation[a]	Speed of Measurement	Limitations (Examples)
Two-dimensional DIGE	Overnight soaking of IPG strips	Approximately 20 hours for first and second dimensions	Image analysis; statistical analysis is not reliable when more than one protein is in one spot
ProteomeLab PF2-D	Usually multistep, time-consuming	Approximately 48 hours, two or three samples a week	Relatively large initial sample required
MudPIT	Digestion of complex mixtures of proteins in solution	Depends on the number of cycles	Alternative to gel electrophoresis; limitations include reproducibility and scalability
Protein arrays	Fast and easy, protocols available	Fast, utilizing technologies already developed for genomic studies	Limited to a number of good-quality antibodies
SELDI-TOF	Fast and easy, robotics also available	Fast	BiomarkerWizard allows fast data analysis but requires additional statistical analysis; shows only differences in profiles

[a]Does not include sample prefractionation.

the two aspects are developing at a different pace. Techniques for high-throughput screening are effective, although usually laborious and long (except for protein arrays). Methods of information processing are improving quickly and constant development of databases can be noticed. What remains behind at this moment is efficient extraction of information from data (e.g., how much information do we miss during interpretation of mass spectra?). Substantial progress in that area will greatly enhance our knowledge.

SEQUENCE COMPARISON AND DATABASE SEARCHES

With the advent of faster and more reliable technology of sequencing nucleic acids and proteins, very large centralized biological databases have become available. The massive size of the biological databases available and the high rate of growth have influenced the types of research conducted currently. Researchers are focusing more than ever on maximizing the use of biological databases. In addition, biology has a long tradition of comparative analysis leading to knowledge and discovery. Hence, it would be very advantageous for bioinformatics researchers to utilize the information stored in current databases to extract new biological information as well as to understand various biological phenomena. Therefore, comparing biological sequences is one of the most important bioinformatics operations. There are several ways to measure the similarity or dissimilarity between two biological sequences, five of which are presented below.

Hamming Distance

The simplest way to compare two given sequences is to count the number of positions they differ in or the number of mismatches in the two sequences. This can be measured using the concept of Hamming distance as developed in the domain of coding theory. The Hamming distance can be viewed as the editing distance between the two sequences. For example, the two sequences AGTTGGA and AGGTAGA differ in two positions, the third and the fifth positions. Hence, the Hamming distance between the two sequences is equal to two. This is rather simple measure of dissimilarity since it fails to deal properly with sequences of different sizes.

Sequence Alignment

Biological sequences do not always have the same length. A robust method for measuring the degree of similarity between sequences has to handle the variability in length properly. Sequence alignment uses the notion of gaps to address this problem. In comparing two sequences, the goal of a sequence alignment algorithm is to insert gaps in the shorter sequence so that the alignment score between the two sequences is maximized. The alignment score is obtained by assigning a score to each position based on whether two characters in the position match, mismatch, or align a character with a space. A typical scoring system assigns +1 for each match, −1 for each mismatch, and –2 for aligning a space with a nonspace. The alignment score is then calculated as the sum of the scores at each position. Consider the two sequences AGTTGGATG and ATGGGTG; alignment algorithms will attempt to insert two spaces in the second sequence such that the total alignment score is maximized. In the alignment shown below, there are six matches, one mismatch, and two spaces, resulting in an alignment score of $6^*(+1) + 1^*(-1) + 2^* (-2) = 1$. Note that this alignment is the optimal alignment for these two sequences.

AGTTGGATG
A - TGGG-TG

The alignment concept abstracts the process of finding out how many steps are needed to convert one sequence to the other, using the three operations on insertions, deletions, and substitutions. There are several variations of the alignment problem, depending on the scoring system used, the way of penalizing gaps, and whether the goal is to align two or more sequences. The latter version is known as the *multiple sequence alignment*. Also, while global alignment attempts to align entire sequences, the goal in local alignment is to align local subsequences within the input sequences. For details on the variations of the alignment problem, see the books by Setubal and Meidanis (1997) and Jones and Pevzner (2004).

Alignment of protein sequences is conducted in a way very similar to aligning DNA sequences. However, a few differences need to be observed. For example, certain amino acid substitutions are commonly observed in similar proteins from different species. Since the protein sequences have functional implications, amino acid substitutions are likely to maintain protein structure and function. Such substitutions result in a chemically similar amino acid, while other substitutions are observed less frequently. Assigning different scores for different amino acid substitutions can be used to reflect the likelihood of such substitutions happening in nature (Durbin et al., 2003).

How can the alignment problem be solved? Although it may appear that finding the optimal alignment can only be achieved using a slow brute-force approach, dynamic programming (DP) can be used to find optimal alignments in a reasonable time. DP is an algorithm technique used to find the best possible solutions for many optimization problems. DP algorithms find an optimal solution by dividing the problem into smaller overlapping subproblems, solving each only once, storing it in a tablelike structure and finally, using the table to construct a solution of the large original problem (Durbin et al., 2003).

Although DP alignment algorithms find the best solutions for any global or local alignment problem, often researchers use BLAST to meet their alignment needs. BLAST (basic local alignment search tool) is a famous alignment heuristic that is probably the most widely used bioinformatics algorithm. Although it does not always find the optimal alignment, its speed has made it the alignment tool of choice in comparing a (new or unknown) sequence against a large set of (known) sequences. BLAST tries to locate a list of short, frequently found subsequences (words) in the new sequence, which is also present in the remainder of the sequences. Then the algorithm extends the common words in both directions as long as the alignment score is acceptable, leading to a possible alignment of the sequences. Unlike DP algorithms, which require quadratic time (in terms of the sequence size), BLAST achieves its goal in linear time. Hence, it sacrifices accuracy in favor of blazing speed.

Advanced Sequence Comparison Methods

Although sequence alignment is the sequence comparison method used the most, there are cases in which alignment fails to measure similarity properly. For example, in the case of subsequence repeats, alignment would incur a heavy penalty since the repeated subsequence would need to be aligned with a gap. Dictionary-based sequence comparison handles this case nicely. In comparing two sequences, a dictionary-based algorithm scans one sequence and constructs a list of unique words, forming a dictionary. The algorithm then scans the second sequence starting with the list constructed. The number of additional words added while scanning the second sequence is used as a measure of dissimilarity. The relative measure similarity (RMS) algorithm uses a data compression technique to construct the dictionary and calculate the score (Otu and Sayood, 2003).

Searching Sequence Databases

The most frequent type of search used in bioinformatics is to compare an input sequence against a certain target database. Such a search is usually conducted to obtain knowledge about the new sequence. For example, it could be used to obtain information regarding protein structure or the function of a gene. Clearly, this process is performed using sequence comparison algorithms. If the target database is large, BLAST is an easy choice. A BLAST search returns a ranked list of sequences according to the degree of similarity to the input sequence.

Another common operation is searching for common motifs in a database of biological sequences. *Motifs* are short sequences with a specific function, such as associating with binding sites. This is often a much harder task since the desired motif is often not known and may not exist as a consensus motif in the target sequences. The quality of a search algorithm is often measured by its sensitivity and selectivity. *Sensitivity* is the parameter that measures an algorithm's ability to find the desired sequences; *selectivity* measures the algorithm's ability to exclude false positives.

Sequence Comparison and Gene Prediction

Recently, due to the advancement of sequencing technology, the sequencing of genomes of many organisms has been completed. However, biological interpretations of these genetic sequences, such as annotation, are not keeping pace with this flood of raw sequenced data. Hence, there is a tremendous need for accurate and fast tools to analyze these sequences and, most important, to find genes and determine their functions (Mathe et al., 2002). Several gene prediction tools are currently available to the public. In general, these tools may be classified in two groups. The intrinsic-based approaches use statistical algorithms to find the optimal parse of a fragment of genome to identify protein-coding and protein-noncoding parts, based on global and local compositional features of specific species. The specificity and sensitivity of predictions by these methods are not always acceptable, especially in the context of long genomic fragments containing multiple genes. Moreover, statistical methods rely on information derived from known genes of specific species, therefore are not useful in the prediction of newly sequenced organisms.

The second group of methods, similarity based, is based on sequence comparison. These approaches are based on a plausible assumption derived from the theory of evolution. During evolution, the functional parts may have encountered more pressure of selection, and they tend to be more conserved than do nonfunctional parts. Thus, when the genomes of different organisms are compared, the local sequence conservation is usually indicative of biological functionality. It can be assumed, in gene-finding problems, that protein-coding regions, as functional parts of sequences, are more conserved than are noncoding regions. The advantage of similarity-based approaches is that they rely on accumulated preexisting biological data and thus produce biologically relevant predictions. However, similarity-based methods tend to be biased toward finding genes that are similar to known genes and cannot find genes encoding new proteins (Batzoglou et al., 2000; Morgenstern et al., 2001).

Since the large fragments of genomic DNA and even complete chromosomes of some eukaryotic organisms are available, it is possible to apply comparative genomics in identification of protein-coding regions (Morgenstern et al., 2001). A gene-finding model may then be developed in which the protein-coding regions can be identified by comparing genomic sequences of two species. Given sequences containing orthologous genes from two related organisms, the most conserved fragments can be obtained using global alignment algorithms. The conserved splice junctions and start/stop codons are identified using local alignment procedures (Chen and Ali, 2005).

DATA ANALYSIS

As mentioned earlier, the advancement in sequencing technology has made it possible to produce a massive amount of biological data in a limited time. While having these data has provided researchers with a valuable source to use to advance their research, it has created the difficult task of how to extract useful knowledge from the raw data collected. Hence, various data analysis techniques need to be used to allow researchers to take advantage of the data. Clustering algorithms, data-mining tools, and statistical systems are popular and useful tools for help in that regard.

Clustering Algorithms

Given an input set of elements, clustering is a traditional algorithmic technique that allows us to hide the fine details in large data systems and helps us to see overall patterns associated with the data. The basic concept in clustering is to group together elements with similar attributes and form clusters of elements. Successful clustering algorithms form groups such that elements in the same group have similar attributes or properties as defined by the user, whereas elements in two different groups possess relatively different attributes. Hence, the success of a clustering algorithm can be measured by the similarity of the elements in the same cluster and the dissimilarity of elements in two different clusters. While grouping a subset of elements under one cluster can be viewed as losing information, it makes it possible to extract patterns and useful knowledge buried in the detailed raw data. With the nature of the massive amount of biological data, clustering is essential to analyze such data. Clustering algorithms have been used frequently and successfully to organize multivariate biological data with similar patterns into distinct groups. They have been applied to address several bioinformatics problems, such as the analysis of gene expression and the construction of phylogenetic trees (Nei and Kumar, 2000; Shamir and Sharan, 2000).

Finding the perfect cluster is a very difficult computational problem. As a result, many heuristics have been developed to cluster data from various application domains. In bioinformatics literature, a number of clustering algorithms have been used, including agglomerative algorithms such as hierarchical clustering, optimization methods such as k-means, graph theoretical–based algorithms such as CLICK and CAST. Refer to the book by Shamir and Sharan (2000) for a description and list of the characteristics of each algorithm and for a comparative review of those methods in an expression profile. While these heuristics have been helpful in the analysis of biological data, the fact that they were not designed for the domain of bioinformatics leaves room for improvements since biological data may have some unique properties. For example, microarray expression profiles may contain a great deal of noise. The clustering algorithm should be able to allow noise features to be suppressed and signal features to be enhanced. Recently, a new clustering algorithm that employs the concept of message passing was introduced. Message passing clustering (MPC) allows data objects to communicate with each other, generates clusters in parallel so that the global distances and local distances are well balanced, and hence improves the performance of clustering. Several experimental studies showed that MPC is particularly effective in analyzing biological data (Geng et al., 2007).

Use of Clustering Algorithms in the Representation of Protein Sequences

One bioinformatics application that takes advantage of clustering techniques is the simplification of the alphabet in the representation of protein sequences. Proteins are generally represented as a sequence of symbols, where each symbol represents one of the protein's amino acids. This set of symbols is known as the *amino acid alphabet*. There are 20 amino acids that can appear in a protein, so there are 20 symbols in the standard amino acid alphabet. For some biological problems it is useful to work with a reduced set of symbols, which is known as using a simplified amino acid alphabet. Grouping two or more amino acids together and representing them by the same symbol (based on some desired biological properties) simplifies the representation of the protein sequences. Simplified alphabets have been used to predict protein interactions, infer protein function, and search for unanticipated patterns in proteins (Cannata et al., 2002; Gomez et al., 2003). A barrier to the efficient creation of simplified alphabets is selecting from a massive number of possible simplifications since there are 51×10^{12} possible simplified alphabets. However, clustering algorithms can be used to find a suitable grouping. Ultimately, the effectiveness of a simplified alphabet depends on its ability to provide insight into a microbiological problem (Palensky and Ali, 2006).

Data Mining for Bioinformatics

Extracting useful knowledge from the massive heap of the biological data available is the most important task in bioinformatics. Although clustering techniques can provide some support in addressing such a difficult problem, it is clear that advanced data-mining tools are necessary to mine for the information hidden in the heap of raw data. Data-mining tools are basically a set of algorithmic techniques and software systems that address the problem of how to extract specific information from large data sets. Bioinformatics is a perfect domain for the utilization of data-mining tools, due to the urgent need to look for knowledge in the available sequences, structures, and microarray data.

Biological data are available in different shapes and forms. The way the data are represented and stored affects significantly the ease and effectiveness of data-mining algorithms. Unfortunately, the majority of public databases were originally designed for archival purposes rather than for data-mining purposes. In addition, the public databases are very heterogeneous in terms of its quality and completeness. As a result, many research groups create their own local data sets for certain organisms by extracting and curating the relevant data from the public databases, augmenting from the data obtained in their labs, as shown in Figure 3. Validation algorithms

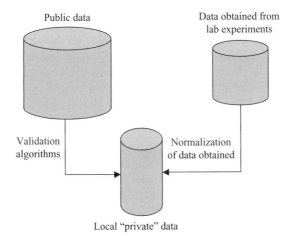

FIGURE 3   Public and private databases.

are necessary to filter the public data extracted and enforce a user-defined level-of-quality measure in order to keep homogeneous custom data for the domain of interest to the lab researchers. Such algorithms may require that the acceptable sequence be complete (contain all components of the sequence) and recent (deposited after a certain date).

There are several publicly available biological data sources. The principal DNA sequence databases are Genbank (http://www.ncbi.nih.gov/Genbank), EMBL (http://www.ebi.ac.uk/embl), and DDBJ (http://www.ddbj.nig.ac.jb). For protein databases, SwissProt (http://www.ebi.ac.uk/swissprot), TrEMBL (http://www.ebi.ac.uk/trembl), and PIR (http://pir.georgetown.edu/pirwww) are the key public sites. Continuous efforts to collaborate and merge various public databases are ongoing to improve the accessibility and usefulness of such critical biological sources.

Data Mining in DNA Databases   Sequence comparison algorithms and clustering techniques are often used for DNA data analysis and to build phylogenetic relationships. Comparing a new sequence to a data set of DNA sequences is a simple approach to determining whether the new sequence belongs to the data set or whether it is similar to other sequences in the known set. Although this may appear to be a simple process, it is not often the case. This is due primarily to the fact that biological sequences are not always deterministic, since positions in the sequences are missing or not uniquely specified. Comparing sequences that are not completely defined is a difficult task. Similarly, looking for sequences that contain a motif (that carries a specific biological function) is not always easy, since it may be composed of several subsequences, some of which are well defined whereas others may not. Exploring ways to represent such nondeterministic sequences and motifs effectively is critical in developing data-mining algorithms in bioinformatics. One way to address the problem of looking for nondeterministic sequences is to use regular expressions. Regular expressions are a robust and powerful tool for representing and manipulating strings composed of a finite set of the alphabet. The regular expression uses the set of alphabet letters, a set of metacharacters, and a finite set of alphabet operators to represent a string or a group of strings with given properties. It employs the concepts of regular grammars to represent strings with a complex set of properties in relatively simple and easy-to-use expressions. For example, the simple regular expression AC(GT)*GA represents a group of DNA sequences that start with the two nucleotide symbols AC and the end of the symbols GA while having any number of GT pairs in between. Several grammar-based algorithms have been developed to search for specific DNA sequences or motifs. Hopefully, with further advancements of sequencing techniques, the degree of nondeterminism in describing biological sequences will be reduced, allowing for faster and more efficient methods for extracting desired knowledge from the data available.

Data Mining in Protein Databases   Protein data analysis can be described as more complex than DNA data analysis. Although proteins are often stored as linear sequences of amino acids, a full understanding of their function requires additional information about their three-dimensional structures. Hence, proteins can be compared in terms of their sequence similarity and/or their structural similarity. Sequence similarity can be addressed in a way analogous to the one used to address similarity between DNA sequences. Primarily, sequence comparison methods (such as alignment algorithms) can be used to measure how similar or dissimilar protein sequences are, and generally, DNA data analysis techniques can be applied to protein data analysis when represented in its linear sequence presentation. The three-dimensional structure can be represented by x-ray crystallography or nuclear magnetic resonance (NMR) spectroscopy. Recently, more protein structures have become available in public databases. Based on the availability of these known structures, computational techniques for predicting protein structures have been developed. These techniques are based on comparing the linear sequence of the protein of an unknown structure with protein sequences of known structure. This is based on the assumption that the more similar the linear sequences of two proteins are, the more similar their three-dimensional structures are likely to be. Additionally, several methods for comparing protein structures have been developed. These methods attempt to measure the structural similarity between the given proteins in terms of the types and arrangements of the structural components of the protein structures such as the α-helices and β-strands. The distance alignment tool (DALI) and vector alignment search tool (VAST) are two popular methods for measuring structural similarities between protein structures (Holm and Sander, 1993; Madej et al., 1995).

Another useful way to represent, compare, and search for proteins is mass spectrometry, which is considerably faster, less expensive, and more accessible than microarrays. As a result, it has been receiving considerable attention lately, especially for clinical applications. In mass spectrometry, proteins are represented by their mass-to-charge ratio *(m/z)* of its chemical components. The output of the mass spectrometer shows a plot of relative intensity versus *m/z*, as shown in Figure 4.

For a given experiment, the raw mass spectra for each sample are composed of features represented by *x, y* pairs, where *x* represents the *m/z* values, with corresponding intensities represented on the *y* axis. Typically, the number of features is much larger

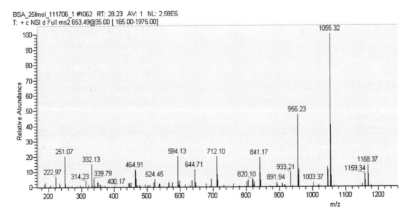

FIGURE 4   Peaks in mass spectrometry.

than the number of samples. Focusing on the peaks in the graph is used to reduce the number of features, with peaks being features with local maximum intensities. Peak detection methods are usually accomplished by the software included with the spectrometer being used. The sequences of peaks of the ratios provide a signature that can be used to identify and compare proteins.

GEL ANALYSIS IN TWO-DIMENSIONAL ELECTROPHORESIS

Three technological advancements elevated two-dimensional electrophoresis (2-DE) to a major technology platform in proteomic profiling: First, the development of immobilized pH gradients increased reproducibility; second, the development of DIGE technology allowed quantitative analysis of protein spots; and third, the development of software for gel analyses. Handling data from 2-DE gel images is a time-consuming process, due to the complexity of 2-DE patterns. It is impossible to perform such an analysis without the help of bioinformatics, which is still a bottleneck in proteomic studies. Very often, a manual adjustment is necessary. Although we observe the constant development of software for 2-DE gel analysis, these tools still experience several shortcomings in methods for protein quantitation and comparisons which are sensitive, rapid, simple, reliable, reproducible, and automated with limited intervention of researcher. Table 2 summarizes available software, although this list may not be complete.

TABLE 2   Available Software for Analysis of 2-DE

Software	Company	Reference
DeCyder 2D	GEHealthcare, Inc.	http://www4.gelifesciences.com/
Delta2D	DECODON, GmbH	http://www.decodon.com/Solutions/ Delta2D.html
ImageMaster 2D Platinium and Melanie	Swiss Institute of Bioinformatics, GeneBio and GEHealthcare	http://ca.expasy.org/melanie/
Dymension	Syngene	http://www.2dymension.com/html/ dymension.html
PDQuest	BioRad Laboratories, Inc.	http://www.biorad.com/
GelFox	Alpha Innotech Corp.	http://www.alphainnotech.com/
ProFINDER 2D	PerkinElmer, Inc.	http://las.perkinelmer.com/
Progenesis SameSpots	Nonlinear Dynamics	http://www.nonlinear.com/
Progenesis 2D	Shimazu Biotech, Inc.	http://www.shimadzu-biotech.net/
HT analyzer	Genomic Solutions	http://www.genomicsolutions.com/
Gellab-II	Image Processing Section of the Laboratory of Experimental and Computational Biology, DBS in the National Cancer Institute/FCRDC	http://www.lecb.ncifcrf.gov/gellab/

For a good comparison, a 2-DE map should show spots that are not overlapping, dark, have well-defined borders, and be free of streaks and background staining. IPG strips with various pH ranges and custom-made gels for the second dimension certainly help to distribute spots equally throughout a gel as well as in areas of interest. Upgraded versions and added functions make software more reliable, reproducible, and automated, as well as help in the analysis of partially separated protein spots. Nevertheless, we have to keep in mind that poor separation, which results in irreproducible gels, cannot be corrected by any software. It has been proposed that replicates with greater than 20 to 30% variability should not be included in the analysis (Goldfarb, 2007). Therefore, proper experimental design and sample preparation are necessary and will determine the quality of data derived from 2-DE experiments as well as the statistical significance of 2-DE analyses. Many criteria, such as normalization, reproducibility, and selection of spots, are very often user defined and as such, may vary from operator to operator within the same laboratory and, more important, between laboratories and centers.

A major challenge in 2-DE gel analysis is gel matching or gel alignment. This is one of the essential, time-consuming prerequisite steps necessary for a quantitative and qualitative analysis of gel images. Ideally, gel alignment is performed in an automated and reliable manner. However, if two spots are too far apart due to the variability of electrophoretic separation (e.g., gel), matching might not be possible, or two unrelated spots could be matched, introducing false positive results. The magnitude of this problem grows as the number of gels used for analysis increases. The same problem applies to the analysis of gels generated in different laboratories.

Spot detection is another issue. Detection of fluorescently labeled proteins is less of a problem because the highly sensitive techniques of scanning, shape, smearing, and streaking introduce great variability into the analysis. More difficulties arise from inherent noise and irregular geometric distortions and partial separation of protein spots. Quantitative analysis fails if there is more than one protein in one spot. In some instances, subsequent Western blot analysis shows that one spot contains one protein that is down-regulated and one protein that is up-regulated. Therefore, software will not replace fuzzy logic thinking by scientists in the interpretation of 2-DE results.

Among many aspects of the quantitative analysis of 2-DE gels, another obstacle is a linear measure of protein abundance over a wide dynamic range. This is important for gel staining as well as gel scanning (Mahon and Dupree, 2001), in particular when proteins are labeled with fluorescent dyes such as Cy3, Cy5, or Cy2. Algorithms define the edge of the spot, which is followed by summing up the intensity of pixels within a defined area. Therefore, laser intensities have to be adjusted so that emission intensities do not exceed the upper limit, and they are prone to operator error. On the other hand, spots with very low abundant proteins and low intensities may not be detected.

Another important function in 2-DE analysis that can introduce errors is proper alignment of multiple gels and background subtraction (Shi et al., 2007). Although algorithms are designed to aid in this process, software users may have an influence on how gels are scanned, thus how high a background an analyzed gel may have. More detailed description of software-induced variance in such analyses may be found in a review paper by Wheelock and Buckpitt (2005). The authors postulate that the software itself constitutes a significant source of the variance observed in quantitative proteomics, and that the use of background subtraction algorithms can increase this variance further.

Variability between gels was another common problem in 2-DE experiments which was diminished by introducing internal standards into 2-DE difference gel electrophoresis (2-D DIGE) (Unlu et al., 1997). Gels using the DIGE method are mixtures of three samples labeled with the three distinct fluorescent dyes Cy2, Cy3, and Cy5. Typically, two dyes are used to label two different biological samples of interest, and the third dye can be used to label the internal standard. It is a pooled mixture of all the samples used in the experiment, and its quantity is identical in each and every gel. The internal standard serves as a reference point for the normalization of intensities, which is expressed as relative abundance.

Summarizing, we will observe containing development in image analysis, spot detection, normalization, automated matching, and quantification which will facilitate the large-scale proteomics experiments necessary for the discovery of statistically significant differences.

THE UNQUESTIONABLE BENEFITS OF SHARING INFORMATION

The tremendous advancement in computational sciences has transformed many sciences, including the biological sciences, not just by providing more data and speeding up steps traditionally done without automation, but by providing novel and powerful perspectives which often lead to unforeseen insights. However, to effectively take advantage of such advancements, new ideas need to be developed and new tools need to be designed. Such new ideas and tools must combine the ability to utilize technological advancements with the ability to take advantage of domain expertise to close the potential gaps between the emerging technology and the application domain.

For example, the recent technological explosion had two major impacts on how scientific studies are conducted. First, through multiplexing and increased throughput, we are now able to perform complex analyses at high speeds and hence generate a tremendous amount of data. Second, posting these data on the Web for global viewing makes it possible to make the data available to almost everyone who has basic computer skills. The availability of such a huge and diverse database provides major opportunities as well as presenting significant challenges in advancing biomedical research.

The universality of the World Wide Web and the fact that it is hypertext based makes it possible to link all types of information with little discrimination between incomplete drafts and well-reviewed documents, between flat files and well-structured databases, or between commercial data with limited access and open academic and governmental information. In many biological domains, the data available are likely to be incompatible with each other in many aspects, including the format of the data sets and their level of reliability. This heterogeneity of Web-based data and the lack of a simple quality control mechanism have limited the tremendous benefit to be gained from such a great resource. This leads to many problems, and attempting to incorporate available data into one model may, in some cases, lead to losing valuable information in the process.

There is a worldwide effort to tackle format-related problems and to make biological data available in a uniform format. Two examples that best illustrate such efforts in the area of genomics and proteomics are the Universal Protein Resource (UniProt) (http://www.ebi.uniprot.org/index.shtml) and World Human Proteome Organization (http://www.hupo.org/) initiatives such as the Human Plasma Protein Project. UniProt is a central repository of protein sequence and function. It is the world's most comprehensive catalog of information on proteins and consists of three components: the UniProt Knowledgebase (UniProtKB), the UniProt Reference Clusters (UniRef), and the UniProt Archive (UniParc). Each component is optimized for different uses. For more detailed descriptions of these components, we refer readers to http://www.ebi.uniprot.org/index.shtml.

Although it is critical to coordinate efforts and standardize the way that biological data are presented, simplifying their use and maximizing their benefits, this is a relatively small problem compared to other problems related to the explosion in the amount of biological data available. In particular, there is an urgent need to develop methods not only to collect but also to filter needed data using user-specified parameters in order to best utilize such data. There are several initiatives in that regard, and although most were not developed exclusively for proteomic and genomic needs, the life sciences are expected to be the primary beneficiary. The development of the concept of a *semantic Web* is a key effort in that regard.

There are many ways to define a semantic Web. Berners-Lee, considered the originator of the concept, defines it as "an extension of the current Web in which information is given well-defined meaning, better enabling computers and people to work in cooperation" (Berners-Lee et al., 2001, 2006; Berners-Lee and Hendler, 2001. The current Web was developed with people in mind and presents information in a natural language–oriented format. Although this is appropriate for human users, it makes it difficult to develop algorithmic tools that utilize the Web effectively since computers communicate best in formal languages. The semantic Web would also allow various groups to develop tools specific to their domain of interest and use their own terminology to collect and filter data they need for their use. Potentially, this will allow researchers to find, share, and combine information much more easily.

To illustrate the benefits of sharing information, we will use an example of fulfilling Koch's postulates for proving that HIV-1 is a causative agent of AIDS. Four criteria postulated by Robert Koch in 1890 are that (1) the microorganism must be found in all cases of the disease, (2) it must be isolated from the host and grown in pure culture, (3) it must reproduce the original disease when introduced into a susceptible host, and (4) it must be found in the experimental host so infected. Fulfilling these four postulates for proving that a given microorganism is the cause of a given disease has some limitations, such as the fact that particular microorganisms (e.g., the one that causes leprosy) cannot be "grown in pure culture" in the laboratory, there is no animal model of infection for a particular microorganism, microorganisms often infect immunocompromised hosts, or a microorganism acquires an extra virulence factor, making it pathogenic. Therefore, for many years, fulfilling Koch's postulates in proving that HIV-1 is a causative agent of AIDS was a problem. Eventually, based on a combination of data and information collected from many studies (both clinical and basic) and many sources, the National Institute of Allergy and Infectious Diseases (NIAID) came to the conclusion that the four criteria had been fulfilled. For more detailed information about this particular case, we refer readers to http://www.niaid.nih.gov/publications/hivaids/hivaids.htm.125%. To reach such a conclusion required exchange not only within the scientific community but also from health reports related to this disease, and there is no doubt that sharing information played a pivotal role in this process.

BRIDGING PROTEOMICS AND GENOMICS

The basic functions of many normal physiological as well as pathological processes are still unclear and pieces of information from genomics and proteomics are still missing. For example, O'Hara and coauthors, studying normal and pathological sleep, conclude that it is more difficult to achieve progress at the protein (proteomics) level than at the genomic level, which uses a more traditional approach, such as abundance of mRNA or a more advanced quantitative trait loci (QTL) approach (O'Hara et al., 2007). Therefore, an ability to combine the benefits of current advancements in proteomics in conjunction with genomics and other technology platforms will transform science and accelerate realization of the promise of answering many questions related to the functioning of entire systems.

It is evident that completion of sequencing the human genome showed a huge gap between the number of genes and the number of existing proteins. Before the final product, a functional protein, is made, intermediates can undergo multiple modifications, such as alternative splicing at the transcription level and over 300 known posttranslational modifications, which can exist in a variety of combinations, fragmenting, cross-linking, and so on (Figure 5). Moreover, open reading frames (ORFs) have the potential to encode a polypeptide (part or entire protein), but many

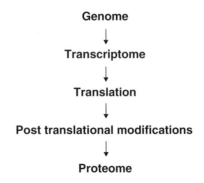

FIGURE 5   Steps leading from genome to proteome.

may not actually do this. To study all of steps presented in Figure 5 requires the use of different techniques; thus, the format of the information output will be different. Thus, the gap between the number of genes and the number of existing proteins will result from possible combinations of all these factors.

An interesting approach in closing this gap has been proposed by Pradet-Balade et al. (2001): using profiling of polysome-bound mRNA. The authors claim that total mRNA does not reflect the translated portion of mRNA and therefore cell phenotype in the first step will depend on translated but not on the total pool of mRNA. Profiling of polysome-bound mRNA also enables researchers to show the difference between transcriptional and translational regulation and generates reproducible and quantitative data (Pradet-Balade et al., 2001). In consequence, the ratio of the level of protein to its corresponding mRNA can be as high as 30-fold, with a lower correlation in the case of less abundant proteins. We expect that many biomarkers are in the category of medium- to low-abundance proteins. It may explain to some extent why it is so difficult to make a connection between results from in vitro and ex vivo studies. Variability in the clearance of proteins from circulation (i.e., CSF, blood, or other body fluids) may increase this disparity. Therefore, understanding of the mechanisms controlling the activity of translation and posttranslational machinery in biological processes is crucial for effective intervention (e.g., in disease management or diminishing developmental deficiencies) (Day and Tuite, 1998; Mathews, 2000).

Additional regulatory mechanisms, such as nuclear transport, stability of transcripts, and protein degradation, add to the complexity and variability of protein isoforms and their levels in the biological system. The functional status of protein(s) is the most difficult problem to harness. For example, the functional diversity of isoforms is generated by posttranslational modifications such as glycosylation, phosphorylation, or proteolytic cleavage. Function(s) not only can be acquired but can also be modified and/or altered at the posttranslational modifications level (Harry et al., 2000). Another level of complexity is the reversible functionality of proteins as a result of PTM, such as phosphorylation. There is no doubt that computer modeling of molecules and their interactions, databases of domains, and consensus sequences are all very helpful, yet it does not give us a precise description of mechanisms and consequences of changes in these mechanisms due to, for example, infection or malignant information. Therefore, we have to expect lengthy and systematic bench experimenting dipping equally into all old and new technologies.

However, this is only a part of a problem because it does not tackle cellular and/or extracellular localization. Few examples include moonlight proteins, which change function depending on localization (Jeffery, 2003). Therefore, how do we connect information obtained from genomic and proteomic experiments for correct and effective biological interpretation? Computational biology with genomic and proteomic networks appears as the most promising approach to integrate a great diversity of types of data and allow analyses with multiple changing factors measured using various techniques (Souchelnytskyi, 2005).

So far, genomic and proteomic experiments are still being designed in somewhat simplistic ways. The major reason for this is a difficulty in integration of data from a variety of technology platforms, already existing information, and a variety of formats, languages, and bioinformatic concepts. Therefore, it is necessary to reduce the number of variabilities as well as to unify data organization to be compared during analysis steps. In a typical genomic/proteomic experiment a comparison is made between two samples (e.g., treated and untreated) or multiple samples from two homogeneous groups of samples (e.g., healthy and diseased people). Although the number of elements (genes or proteins) in such comparisons can be counted in the thousands, it still remains as a snapshot rather than evaluation of the dynamic biological process. The dynamic complexity of any living organism, even one with a relatively small genome, such as viruses, is just one example of a limitation. Infection of a T-cell or macrophage with HIV-1 has profound global effects which are not limited to the virus–host cell interaction itself but also relate to opportunistic infections, neurodegenerative disorder, and so on. There is hope and expectation that proteomics as a complement to genomics revolutionizes screening, which will have a great impact on understanding how biological systems work (Bahtiyar et al., 2007), and more results of such studies are being published (Titball and Petrosino, 2007).

BIOSTATISTICS AND DATA ANALYSIS

Many issues in bioinformatics are addressed by collecting data and exploring potential relationships among the data elements collected. It is often very difficult to use deterministic techniques to approach such situations—hence the need for biostatistics. Biostatistics is the scientific field that deals with the development and application of statistical models and techniques to provide the much needed stochastic data analysis in biosciences.

Since the explosion of data available to researchers, biostatistics has become an indispensable tool in extracting useful knowledge from raw biological data. Biostatistical reasoning and modeling has been critical to formation of the theoretical foundation of modern biology. In addition to data analysis, biostatistics plays an important role in designing experiments in which data collection is used to infer potential relationships. This includes formulating the scientific questions to be answered, determining the appropriate sampling techniques, coordinating data collection procedures, and carrying out statistical analyses to answer those scientific questions. Examples of such studies include developments in likelihood methods for inference, epidemiologic statistics, clinical trials, survival analysis, and statistical genetics.

Major problems in biomedical research have sparked the development of advanced statistical methods, which in turn have improved the ability to draw potential correlations and inferences from available data. A key objective of biostatistics is to advance statistical science and its application to problems in biosciences with the ultimate goal of having a positive impact on the public's health. For example, Bayesian networks have received significant attention from researchers as a powerful stochastic modeling technique, due to their applicability in the bioinformatics domain.

A Bayesian network is a graph-based model for representing probabilistic relationships between random variables. The random variables, which may represent source data such as gene expression levels, are modeled as graph nodes, and probabilistic relationships are captured by directed edges between the nodes. Bayesian inference is the process of fitting a probability model to the input set of data and inferring possible relationships using a conditional probability distribution on the parameters of the model. Bayesian networks have been used to solve many data-mining tasks in bioinformatics, such as classification and diagnosis.

For a comprehensive coverage of biostatistics, refer to the *Encyclopedia of Biostatistics* (http://eu.wiley.com/legacy/wileychi/eob/).

CHEMOMETRICS

The suffix *-omics* refers to studies of groups or systems of biomolecules and their relationships. Consequently, genomics is the study of an organism's genome, and proteomics is the study of an organism's proteins. The inherent property of results generated in such studies, which involve multiple measurements of many samples, consisting of hundreds or even thousands of elements, is the multivariate character of data. Unlike in traditional experiments in which one or two variables are measured at one time, -omics experiments deal with multiple changes of parameters, such as gene expression or protein levels. Usually, the major goal of these experiments is to find patterns of changes in correlation with biological characteristics of the system in question. This approach was used successfully by Strauss and Falkow to identify candidate genes important for virulence in *Haemophilus influenza*. The authors further confirmed the computational data experimentally (Strauss and Falkow, 1997). Chemometrics, the use of mathematical and statistical methods, helps in the design and optimization of experimental parameters, calibration and normalization, pattern recognition, and statistical analysis. This short list certainly does not exhaust all benefits which chemometrics as an analytical tool can provide in building a bridge between methods and their application. Moreover, chemometrics and statistics can also help in evaluation of data already reported during the peer review processing of manuscripts. Chemometrics and statistical tools can evaluate whether conclusions presented by the authors of work submitted for publication are justified based on experimental design and data processing and whether additional data are required (Spiegelman et al., 2006).

Chemometric tools helped in addressing issues in acquisition and analysis of mass spectrometry data. This includes not only peak detection but also baseline subtraction, calibration, and normalization. Another very important function of chemometrics in mass spectrometry data analysis is peak alignment (De Braekeleer et al., 2000). The success of biomarker discovery and the elimination of false-positive results will be proportional to the correct alignment of peaks across thousands of spectra. Chemometric analysis can also identify possible outliers and indicate whether there are patterns or trends in the data set and whether by reducing information to a more comprehensive mode, patterns can be uncovered (Eriksson et al., 2004). This can be accomplished by applying exploratory algorithms such as principal component analysis and hierarchical cluster analysis.

For further studies related to application of chemometrics to proteomic experiments, the authors recommend several sources, such as the *Journal of Chemometrics* and other resources available online: www.spectroscopyNOW.com.

Concluding Remarks

The proteomics boom continues although the objectives of this field are changing from simple cataloging to functional studies of systems biology. Concurrently, bioinformatics as a scientific discipline is also exploding. This is reflected by the steadily increasing number of widely used software tools as well as peer-reviewed publications. Nevertheless, major challenges remain to be faced, including processing and understanding data from high-throughput experiments, integration of information accumulated from various data sources, technological platforms, methods of data acquisition, and storage and presentation of biological data. In the foreseeable future we can expect further rapid development of bioinformatics methods of data analysis and increase in use of these methods at multiple levels. One example is the transformation of the journal *Computers and Biomedical Research* to *The Journal of Biomedical Informatics*. E. H. Shortliffe, editor-in-chief, explains this decision by saying: "In particular, we wish to emphasize papers that elucidate methodologies that generalize across biomedical domains and that help to form the scientific basis for the field." *Briefings in Functional Genomics and Proteomics* positions itself as an international forum of discussion techniques, protocols, and approaches in genomics and proteomics. Certainly everybody will benefit from such resource and they would be indispensable to scientists new to this field. Despite limitations and bottlenecks in the present and future, functional studies such as proteomics and genomics will advance our knowledge, ultimately translating to better diagnosis and treatment of diseases, and to better understanding our surrounding environment. Such efforts will be boosted by further advances in bioinformatics research and by crossing the bridges connecting biosciences and computational sciences.

SELF-STUDY QUESTIONS

1. Sequence alignment has been the method of choice for many researchers to compare biological sequences. However, there are cases for which dictionary-based sequence comparison should be used instead of sequence alignment to compare sequences. Suggest cases for which sequence alignment would be the better method for comparing sequences and cases for which dictionary-based methods would be the better choice.

2. Many genes are shared among different species, and hence, comparing the genomes of two species can be used as a gene-finding approach:
    a. Suggest the steps of a method that uses comparative genomics for finding genes.
    b. What properties would you suggest we use to select the species on which comparative genomic methods can be used for finding genes?

3. Discuss the pros and cons of using public biological data for research purposes. Identify the challenges that researchers face in using public databases.
4. Recognition of proteins can be accomplished using various methods. Compare the advantages and disadvantages of using amino acid motif sequences and mass spectrometry peak sequences in recognizing a given protein sequence.
5. Define the term *systems biology*.
6. It has been found that there are many more proteins than genes in a human genome, which is in disagreement with theory of "one gene, one protein." Discuss reasons for this discrepancy.
7. What is a difference between multiplexing and high throughput? Discuss the importance of these two approaches in proteomic and genomic studies.
8. Why is bridging genomics and proteomics such a highly challenging task? Discuss the role of transformation of data to information to knowledge in this process.

RECOMMENDED READING

Bahtiyar M.O., Copel J.A., Mahoney M.J., Buhimschi I.A., Buhimschi C.S. Proteomics: a novel methodology to complement prenatal diagnosis of chromosomal abnormalities and inherited human diseases. *Am. J. Perinatol.* 24(2007) 167–181.

Barritault D., Expert-Bezancon A., Milet M., Hayes D.H. Inexpensive and easily built small scale 2D electrophoresis equipment. *Anal. Biochem.* 70(1976) 600–611.

Batzoglou S., Pachter L., Mesirov J.P., Berger B., Lander E.S. Human and mouse gene structure: comparative analysis and application to exon prediction. *Genome Res.* 10(2000) 950–958.

Berners-Lee T., Hall W., Hendler J., Shadbolt N., Weitzner D.J. Computer science: creating a science of the Web. *Science* 313(2006) 769–771.

Berners-Lee T., Hendler J. Publishing on the semantic Web. *Nature* 410(2001) 1023–1024.

Berners-Lee T., Hendler J., Lassila O. The semantic Web. http://ScientificAmerican.com, 2001.

Cannata N., Toppo S., Romualdi C., Valle G. Simplifying amino acid alphabets by means of a branch and bound algorithm and substitution matrices. *Bioinformatics* 18(2002) 1102–1108.

Chalmers M.J., Mackay C.L., Hendrickson C.L., et al. Combined top-down and bottom-up mass spectrometric approach to characterization of biomarkers for renal disease. *Anal. Chem.* 77(2005) 7163–7171.

Chen R., Ali H. A new approach for gene prediction using comparative sequence analysis. Presented at the 20th Annual ACM Symposium Applied Computing (SAC'05), Bioinformatics Track, Santa Fe, NM, Mar. 13–17, 2005.

Day D.A., Tuite M.F. Post-transcriptional gene regulatory mechanisms in eukaryotes: an overview. *J. Endocrinol.* 157(1998) 361–371.

De Braekeleer K., Torres-Lapasío J.R., Massart D.L. Improved purity assessment of high-performance liquid chromatography diode array detection data for overcoming the presence of the non-linearity artefact. *Chemom. Intell. Lab. Syst.* 52(2000) 45–59.

Durbin R., Eddy S., Krogh A., Mitchison G. *Biological Sequence Analysis: Probabilistic Models of Protein and Nuclic Acids*. Cambridge University Press, New York, 2003.

Eriksson L., Antti H., Gottfries J., et al. Using chemometrics for navigating in the large data sets of genomics, proteomics, and metabonomics (gpm). *Anal. Bioanal. Chem.* 380(2004) 419–429.

Gelfi C., Righetti P.G. Preparative isoelectric focusing in immobilized pH gradients: II. A case report. *J. Biochem. Biophys. Methods* 8(1983) 157–172.

Geng H., Deng X., Ali H. MPC: a knowledge-based framework for clustering under biological constraints. *Int. J. Data Mining Bioinf.* 2(2007).

Gianazza E. Isoelectric focusing as a tool for the investigation of post-translational processing and chemical modifications of proteins. *J. Chromatogr. A* 705(1995) 67–87.

Goldfarb M. Computer analysis of two-dimensional gels. *J. Biomol. Tech.* 18(2007) 143–146.

Gomez S.M., Noble W.S., Rzhetsky A. Learning to predict protein–protein interactions from protein sequences. *Bioinformatics* 19(2003) 1875–1881.

Harry J.L., Wilkins M.R., Herbert B.R., Packer N.H., Gooley A.A., Williams K.L. Proteomics: capacity versus utility. *Electrophoresis* 21(2000) 1071–1081.

Hogan J.M., Higdon R., Kolker E. Experimental standards for high-throughput proteomics. *Omics* 10(2006) 152–157.

Holm L., Sander C. Protein structure comparison by alignment of distance matrices. *J. Mol. Biol.* 233(1993) 123–138.

Jeffery C.J. Moonlighting proteins: old proteins learning new tricks. *Trends Genet.* 19(2003) 415–417.

Jones N., Pevzner P. *An Introduction to Bioinformatics Algorithms*. MIT Press, Cambridge, MA, 2004.

Madej T., Gibrat J.F., Bryant S.H. Threading a database of protein cores. *Proteins* 23(1995) 356–369.

Mahon P., Dupree P. Quantitative and reproducible two-dimensional gel analysis using Phoretix 2D Full. *Electrophoresis* 22(2001) 2075–2085.

Marshall A.G., Hendrickson C.L., Jackson G.S. Fourier transform ion cyclotron resonance mass spectrometry: a primer. *Mass Spectrom. Rev.* 17(1998) 1–35.

Mathe C., Sagot M.F., Schiex T., Rouze P. Current methods of gene prediction, their strengths and weaknesses. *Nucleic Acids Res.* 30(2002) 4103–4117.

Mathews M.B. *Origins and Principles of Translational Control*. Cold Spring Harbor Laboratory Press, Woodbury, NY, 2000, pp. 1–31.

McMullan D., Canaves J.M., Quijano K., et al. High-throughput protein production for x-ray crystallography and use of size exclusion chromatography to validate or refute computational biological unit predictions. *J. Struct. Funct. Genom.* 6(2005) 135–141.

Millea K.M., Krull I.S., Cohen S.A., Gebler J.C., Berger S.J. Integration of multidimensional chromatographic protein separations with a combined "top-down" and "bottom-up" proteomic strategy. *J. Proteome Res.* 5(2006) 135–146.

Morgenstern B., Rinner O., Abdeddaïm S., et al. Exon prediction by comparative sequence analysis. In *Proceedings of the Human Genome Meeting*, Edinburgh, UK, 2001, pp. 146–147.

Nei M., Kumar S. *Molecular Evolution and Phylogenetics*. Oxford University Press, New York, 2000.

O'Hara B.F., Ding J., Bernat R.L., Franken P. Genomic and proteomic approaches towards an understanding of sleep. *CNS Neurol. Disord. Drug Targets* 6(2007) 71–81.

Otu H.H., Sayood K. A new sequence distance measure for phylogenetic tree construction. *Bioinformatics* 19(2003) 2122–2130.

Palensky M., Ali H. A genetic algorithm for simplifying amino acid alphabet in bioinformatics applications. Presented at the International Conference on Artificial Intelligence and Application (AIA'06), Innsbruck, Austria, 2006.

Pradet-Balade B., Boulme F., Beug H., Mullner E.W., Garcia-Sanz J.A. Translation control: bridging the gap between genomics and proteomics? *Trends Biochem. Sci.* 26(2001) 225–229.

Setubal J., Meidanis J. *Introduction to Computational Molecular Biology*. PWS Publishing, Boston, 1997.

Shamir R., Sharan R. *Algorithmic Approaches to Clustering Gene Expression Data*. MIT Press, Cambridge, MA, (2000), pp. 269–300.

Shi G., Jiang T., Zhu W., Liu B., Zhao H. Alignment of two-dimensional electrophoresis gels. *Biochem. Biophys. Res. Commun.* 357(2007) 427–432.

Souchelnytskyi S. Bridging proteomics and systems biology: What are the roads to be traveled? *Proteomics* 5(2005) 4123–4137.

Spiegelman C.H., Pfeiffer R., Gail M. Using chemometrics and statistics to improve proteomics biomarker discovery. *J. Proteome Res.* 5(2006) 461–462.

Strauss E.J., Falkow S. Microbial pathogenesis: genomics and beyond. *Science* 276(1997) 707–712.

Titball R.W., Petrosino J.F. *Francisella tularensis* genomics and proteomics. *Ann. N. Y. Acad. Sci.* (2007).

Unlu M., Morgan M.E., Minden J.S. Difference gel electrophoresis: a single gel method for detecting changes in protein extracts. *Electrophoresis* 18(1997) 2071–2077.

Wheelock A.M., Buckpitt A.R. Software-induced variance in two-dimensional gel electrophoresis image analysis. *Electrophoresis* 26(2005) 4508–4520.

APPENDIX A

LABORATORY EXERCISES

Anna Drabik, Anna Bierczynska-Krzysik,
Anna Bodzon-Kulakowska, and Adam Moszczynski

SAFETY PRECAUTIONS

- Follow specific safety precautions that are valid in the particular laboratory. This may vary depending on the character of the laboratory (chemical, biochemical, animal house, etc.) and on particular regulations issued by the staff.
- Never enter the lab room or start the work unless it has been approved by the instructor.
- Food, drink, chewing gum, and the like are banned in the laboratory.
- Always pay attention to what you are doing. Behave in a responsible manner.
- Check for the location of safety equipment before beginning work. Learn all regulations, focusing on those devoted to emergency situations.
- Read all manuals carefully before turning instruments on or off.
- All instruments that require high voltage for operation (e.g., mass spectrometers, capillary electrophoresis, polyacrylamide gel electrophoresis) should be operated only in the presence of the instructor.
- Any accident or unusual situation (e.g., spills, injuries, improper functioning of an instrument) should be reported to the instructor immediately.
- Always wear safety glasses, goggles, or other eye protection wear all the time you are in the lab. Disposable gloves should be used whenever necessary. Insulating gloves should be used when searching for reagents in freezers.

Proteomics: Introduction to Methods and Applications, Edited by Agnieszka Kraj and Jerzy Silberring
Copyright © 2008 John Wiley & Sons, Inc.

- Do not touch, smell, or taste any chemicals.
- If a chemical should splash in your eye, immediately flush with running water for at least 20 minutes. Inform the instructor immediately.
- Carefully read the labels on bottles, jars, and tubes, and ask for more specific information whenever necessary and in all cases when you are uncertain.
- Gases and gas tubes should be used only with the instructor's permission.
- Never use chemicals from unlabeled containers, no matter how sure you are that they contain what they should.
- Burners, holders, racks, and other devices may be hot. Your colleagues may not be aware of that. Never leave hot equipment unattended. Never touch any equipment unless you are sure that it is safe.
- All waste should be disposed of properly. This depends on the nature of the waste (i.e., hazardous, nonhazardous, biological, infected, chemical, syringes, needles, animals, etc.). Always consult your instructor as to how to proceed.
- Do not return unused chemicals to their original containers.
- Do not remove chemicals from the laboratory.
- **Always follow the golden rule: Even if you are absolutely sure what to do, consult your instructor anyway!**

Caution: All procedures concerning human and animal handling follow specific regulations. Working with the tissues or fluids already extracted may also require approval. Consult your clinic (human samples), the local animal facility, and the Ethics Committee (all samples).

The laboratory exercises listed on the following pages have been prepared for the specific requirements and instrumentation available in our laboratory. Adaptation of these experiments is easy, and other types of HPLC systems or mass spectrometers can be applied without modifications.

1. POLYACRYLAMIDE GEL ELECTROPHORESIS UNDER DENATURING CONDITIONS

Objectives: To gain knowledge of the preparation of polyacrylamide gels for protein separation, staining, and molecular mass calculations. Also, safety requirements during handling of hazardous material are discussed.

Topics to Prepare: SDS-PAGE electrophoresis (principles of in-gel separation of proteins, buffers, Tris-tricine/glycine buffer, gel staining methods, and applications and limitations of SDS-PAGE) and one- and two-dimensional gel electrophoresis.

Reagents: SDS, glycerol, TEMED, ammonium persulfate, acrylamide, bisacrylamide, Tris-HCl buffer, Coomassie Brilliant Blue, sample buffer, water, methanol, protein standards in water (6 μg/10 μL each): bovine albumin, chicken egg lysozyme, horse myoglobin, horse cytochrome *c*.

Equipment: Mini-PROTEAN III gel electrophoresis system (from BioRad). The complete operation manual can be found at http://www.bio-rad.com/cmc_upload/ Literature/44432/4006157B.pdf.

Other Equipment: Automatic pipettes, multichannel pipettor, centrifuge, vortex mixer, water bath.

Precautions: Carefully clean glass plates to remove fat (Mini-PROTEAN) before an experiment. Follow safety rules and all precautions for working with high voltage. During weighing and preparation of an acrylamide solution, follow the general precautions (lab wear, glasses, disposable gloves). Acrylamide is toxic; use a disposable mouth mask during weighing.

Procedure

Buffer Preparation

- *Electrode buffer* + (*lower*): 0.2 M Tris-HCl, pH 8.9 (24.22 g/L)
- *Electrode buffer* − (*upper*): 0.1 M Tris (12.2 g/L) + 0.1 M tricine (17.9 g/L) + 0.1% SDS
- *Solution A*: 3 M Tris-HCl, pH 8.45 (72.7 g/200 mL) + 0.3% SDS (0.6 g/200 mL)
- *Solution B*: 48 g acrylamide + 1.5 g bisacrylamide/100 mL water
- *Solution C*: freshly prepared 10% water solution of ammonium persulfate

Gel Preparation

- *Stacking gel*: 0.3 mL B + 0.93A + 2.52 mL of water + 30 μL C +3 μL TEMED
- *Resolving gel 1*: 1.52 mL B + 2.5 mL A + 3.48 mL water + 25 μL C + 2.5 μL TEMED
- *Resolving gel 2*: 1.83 mL B + 3 mL A + 1 mL glycerol + 3.24 mL water + 45 μL C + 4.5 μL TEMED

1. Gel for protein separation (100 to 10 kDa) should be prepared as follows: 6/10 resolving gel 2 + 3/10 resolving gel 1 + 1/10 stacking gel (ratios are given starting from the bottom).
2. Prepare the gels.
   Caution: TEMED and ammonium persulfate solutions should be added at the end, just before pouring liquid gel between plates. To accelerate polymerization, you can degas the prepared solutions.
3. Lay resolving gel 1 on resolving gel 2 before polymerization. Pour some water on the gel surface and allow it to polymerize. Remove water with filter paper and overlay the polymerized gel with stacking gel with a comb inserted.

Sample Preparation

1. Sample buffer preparation: 125 mM Tris-HCl, pH 6.8 + 4% SDS + 20% v/v glycerol + 50 mg/mL dithiotreitol (DTT) (or without DTT for nonreductive conditions) + a minute amount of Coomassie Brilliant Blue G-250 until a light blue color appears.
   Caution: Use CBB G. Do not use CBB R.
2. Mix buffer and protein solution at 1:1 v/v ratio and heat it in a water bath for 5 minutes.
3. Apply samples and standards to the gel. Separate until the stain reaches the end of the gel (ca. 1 cm from the bottom).

Separation Settings and Gel Handling

1. Perform the separation step at 30 V when a sample is in the stacking gel, increase to 50 V when in resolving gel 1, and increase to 80 V until it reaches the end of the gel.
2. Fix the gel for about 30 minutes in 50% methanol + 10% acetic acid and stain it for 2 hours in 0.025% CBB-R-250 (10% acetic acid solution). Destaining is carried out in 10% acetic acid.

Preparation of the Report

1. Describe the principle of protein separation using SDS-PAGE and describe a role of the reagents used for gel and buffer preparation.
2. Describe your laboratory procedure briefly.
3. Based on the electrophoretic mobility of known proteins, prepare a calibration curve [dependence between the relative mobility of proteins (the quotient of the process followed by the protein and that followed by the stain) and the logarithm of their masses].
4. Based on the calibration curve prepared, estimate the mass of the protein separated.

2. PREPARATION OF IN-GEL SEPARATED PROTEINS FOR MASS SPECTROMETRY

Objectives: To teach students the practical knowledge of sample preparation for mass spectrometry analysis, including excision of protein spots, in-gel trypsin digestion, micro-desalting, and preconcentration procedures.

Topics to Prepare

- *Proteomics:* definition, basic principles, proteome vs. genome, general analytical procedures, preparation of biological material for analysis

- *Mass spectrometry*: basic definitions of resolution, ion sources, deconvolution, multiple ionization, isotope peaks, analyzers, spectra interpretation

Equipment: Automatic pipettes, centrifuge, vortex mixer (for Eppendorf tubes), water bath, ultrasonic bath.

Precautions: Avoid contamination of samples, reagents, equipment, and so on, with keratins, which are abundant in human skin and dust and may obscure the results.

Solutions Required for This Work

- 100 mM NH_4HCO_3 (400 mg NH_4HCO_3/50 mL H_2O)
- 10 mM DTT (1.5 mg DTT/1 mL 100 mM NH_4HCO_3)
- 55 mM iodoacetamide (10 mg iodoacetamide/1 mL 100 mM NH_4HCO_3)
- 100 mM $CaCl_2$ (11 mg $CaCl_2$ /1 mL H_2O)
- *Solution for digestion*: 50 μL H_2O, 50 μL 100 mM NH_4HCO_3, 5 μL 100 mM $CaCl_2$, 15 μL 0.1 μg/μL trypsin solution in 1 mM HCl. Keep the solution on ice for no longer than 2 hours. Proteomics-grade trypsin is recommended.
- *Buffer for digestion*: 50 μL H_2O, 50 μL 10 mM NH_4HCO_3, 5 μL 100 mM $CaCl_2$.

Other Reagents: Water, acetonitrile, 5% HCOOH.

Procedure

Excision of Protein Bands from Polyacrylamide Gel

1. Excise the band selected (it should contain only a minimal amount of the surrounding gel) using a clean scalpel blade.
2. Cut the gel piece into small squares (ca. 1 × 1 mm).
3. Transfer into siliconized Eppendorf tubes (0.5 mL capacity).
4. Wash the bands with water (ca. 100 μL), vortex, centrifuge, and remove the water.

Destaining

1. Use 100 μL of 100 mM NH_4HCO_3 to cover gel particles; after 10 minutes add an equal volume of acetonitrile, vortex for 10 minutes, place tubes in a water bath for 10 minutes at 37°C, and vortex again.
2. Centrifuge gel pieces, remove the supernatant, and add 100 μL of acetonitrile while shaking the tube gently until the gel particles shrink, become white, and adhere to each other.
3. Remove acetonitrile and dry the gel in a vacuum centrifuge.

Reduction and Alkylation

1. Immerse dried gel pieces with 10 mM DTT/100 mM NH_4HCO_3 (they should be covered with solution) and incubate for 1 hour at 60°C.
2. Cool down tubes to ambient temperature, remove the remaining solution, and cover the gel pieces with 55 mM iodoacetamide/100 mM $NH_4HCO_{3.}$
3. Incubate for 45 minutes at room temperature in the dark.
4. Remove the iodoacetamide solution.
5. Wash the gel particles with 100 μL of NH_4HCO_3 under gentle shaking.
6. Centrifuge the gel and remove the supernatant.
7. Shrink the gel in 100 μL of acetonitrile.
8. Remove the acetonitrile and dry the gel in a vacuum centrifuge.

In-Gel Digestion

1. Cover the gel particles with the digestion solution. The gel pieces should be covered completely. Carry out the reaction for 45 minutes. at 4°C, on ice).
2. After the first 15 to 20 minutes check the samples and add the digestion solution if it is absorbed by the gel.
3. Remove the remaining solution and cover the gel with the digestion buffer.
4. Carry out the digestion overnight at 37°C.

Peptide Extraction

1. Withdraw the supernatant and keep it, as it contains peptides released during overnight digestion.
2. Cover the gel with 50 mM NH_4HCO_3, incubate for 15 minutes at 37°C, and vortex every 5 minutes.
3. Remove the solution and combine with the preceding one.
4. Perform the extraction using 5% HCOOH in acetonitrile (50 : 50) and incubate the solution in a water bath for 30 minutes at 30°C (vortex every 10 minutes).
5. Repeat the extraction step in 5% HCOOH in acetonitrile (50 : 50) and incubate the solution at 30°C in an ultrasonic bath:
   a. Pool together ALL solutions containing extracted peptides.
   b. Dry the extract in a vacuum centrifuge.
   c. Dissolve peptides in 5 μL H_2O/0.1% HCOOH.

The solution prepared containing a peptide map of the protein selected (gel spot) is suitable for analysis by mass spectrometry (MALDI-TOF or nano-ESI-LC-MS).
    More details are available on the following web sites:

http://www.promega.com/tbs/tb512/tb512.pdf

http://donatello.ucsf.edu/ingel.html

http://www.garvan.unsw.edu.au/public/corthals/#protocols

Preparation of the Report

1. Discuss briefly the main phases of your work (explain why you perform particular steps).
2. Suggest a protocol for phosphoprotein identification and localization of phosphorylation sites.

3. BIOINFORMATIC TOOLS FOR PROTEIN IDENTIFICATION

Objectives: Identification of a protein on the basis of its peptide maps.

Topics to Prepare: Proteomic databases, software in proteomics, applications, bioinformatic tools (MASCOT, peptide maps, protein modifications, proteolytic enzymes, cleavage sites).

Materials: Mass spectra of peptide maps recorded in Exercise 2 (preparation of the in-gel separated proteins for mass spectrometry).

Procedure: Figure A.1 presents an example of the form that is to be filled in to identify a selected protein based on its peptide map, using a MASCOT server (www.matrixscience.com) and a SwissProt database. In the *Taxonomy* window, select the *all entries* option to find all traces of human keratins, trypsinogen, and trypsin autolysis products. Although these signals generate some information noise, they are useful for instrument calibration. Choose *trypsin* (the enzyme used in your experiment) and specify the possible number of missed cleavage sites. Consider possible protein modifications, such as methionine oxidation. (Other modifications?) Set the *m/z* error tolerance to 0.6 Da. In the *Query* form, type in *m/z* values of the peaks present on the spectrum (separate values with spaces). This data set can also be entered as a file in the *Data File* field. Start the program using the search button.

Figure A.2 presents the search results obtained for digested protein. The protein selected with the highest probability is disulfide isomerase A3 precursor PDA3 (66% amino acid sequence covered). The sequences of other proteins on the list do not match well with the peptide map. Also, the molecular mass of the isomerase, obtained from the calibration curve, fits well with the mass calculated for the protein.

Figure A.3 shows that three peptides contain oxidized methionine. The majority of fragments contain one or zero missing cleavages. The biggest difference between calculated and experimental masses was 0.17 Da. Figure A.4 shows the sequence of protein identified.

Searching the database on the ExPASy server (www.expasy.org) in the PeptIdent program is performed in a way similar to the preceding approach. Here it is necessary

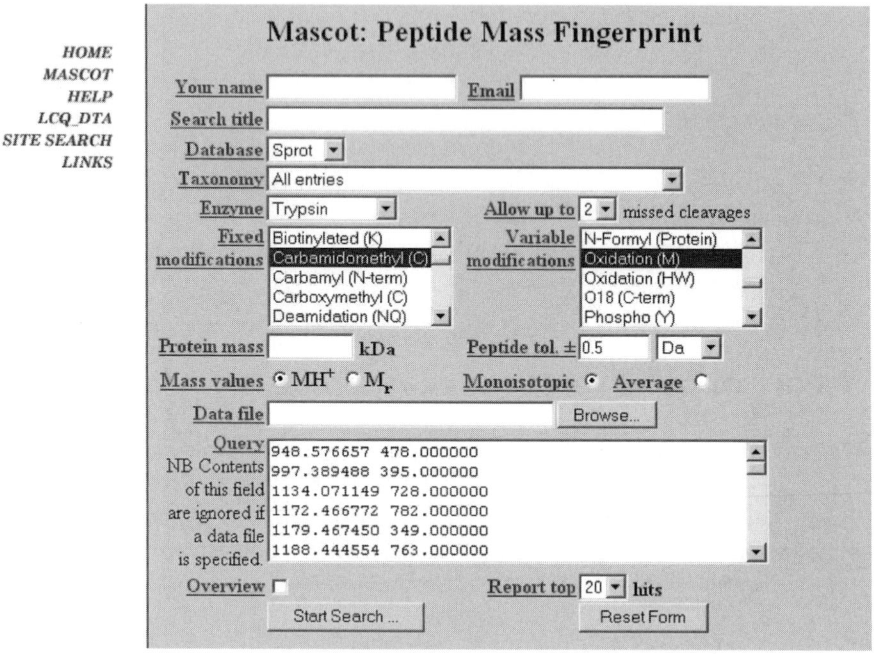

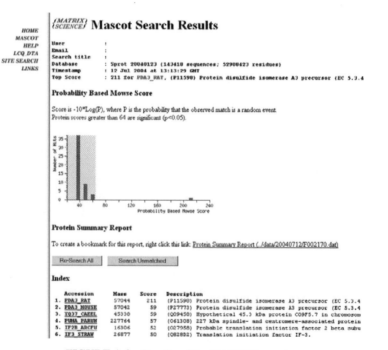

FIGURE A.1   Peptide mass fingerprint program form on the MASCOT server.

FIGURE A.2   Results of the MASCOT search.

1. PDA3_RAT Mass: 57044 Score: 211
(P11598) Protein disulfide isomerase A3 precursor (EC 5.3.4

Observed	Mr(expt)	Mr(calc)	Delta	Start	-	End	Miss	Peptide	
948.58	947.57	947.50	0.06	281	-	288	1	NFVENVAK	
997.39	996.38	996.50	-0.12	153	-	161	0	DASVVGFFR	
1172.47	1171.46	1171.53	-0.07	336	-	344	0	FVNQEEFSR	
1179.47	1178.46	1178.58	-0.12	174	-	183	1	AASNLRDNYR	
1188.44	1187.44	1187.53	-0.09	336	-	344	0	FVNQEEFSR	1 Oxidation (M)
1191.61	1190.60	1190.59	0.01	63	-	73	0	LAPEYEAAATR	
1194.64	1193.63	1193.67	-0.04	416	-	425	2	YKELGEKLSK	
1236.47	1235.46	1235.51	-0.05	108	-	119	0	DGEEAGAYDGPR	
1244.58	1243.57	1243.66	-0.09	184	-	194	0	FAHTNVESLVK	
1341.60	1340.60	1340.68	-0.08	449	-	460	0	GFPTIYFSPANK	
1342.62	1341.61	1341.68	-0.07	215	-	225	1	FEDKIVAYTEK	
1373.60	1372.59	1372.67	-0.07	352	-	362	0	FLQEYFDGNLK	
1394.60	1393.59	1393.65	-0.06	162	-	173	0	DLFSDGHSEFLK	
1395.65	1394.64	1394.68	-0.04	272	-	282	2	NTKGSNYWRNR	
1396.56	1395.56	1395.69	-0.13	367	-	379	0	SEPIPETNEGPVK	
1397.62	1396.61	1396.60	0.01	83	-	94	0	VDCTANTNTCNK	
1469.65	1468.64	1468.77	-0.13	449	-	461	1	GFPTIYFSPANKK	
1488.68	1487.68	1487.77	0.00	336	-	347	1	FVNQEEFSRDGK	1 Oxidation (M)
1529.70	1528.69	1528.77	-0.08	352	-	363	1	FLQEYFDGNLKR	
1587.74	1586.73	1586.81	-0.08	148	-	161	1	FISDKDASVVGFFR	
1588.74	1587.73	1587.87	-0.14	62	-	75	2	RLAPEYEAAATRLK	
1652.69	1651.68	1651.74	-0.06	434	-	448	0	NDATANDVPSPYEVK	
1715.87	1714.86	1714.90	-0.04	147	-	161	2	KFISDKDASVVGFFR	
1721.79	1720.78	1720.94	-0.16	483	-	497	1	EATNPPIIQEEKPKK	
1744.91	1743.90	1743.88	0.02	131	-	146	1	QAGPASVPLRTEDEFK	1 Oxidation (M)
1746.86	1745.86	1745.92	-0.07	289	-	304	1	TFLDAGHKLNFAVASR	
1749.92	1748.91	1748.85	0.06	380	-	395	0	VVVAESFDDIVNAEDK	
1792.73	1791.72	1791.89	-0.17	180	-	194	1	DNYPFAHTNVESLVK	
1801.06	1800.05	1799.93	0.12	364	-	379	1	YLKSEPIPETNEGPVK	
1872.92	1871.91	1871.97	-0.06	131	-	147	2	QAGPASVPLRTEDEFKK	
2302.08	2301.07	2301.14	-0.07	195	-	214	0	EYDDNGEGIIIFRPLHLDK	
2605.25	2604.25	2604.30	-0.05	253	-	274	2	DLIGGKDLLIAYYDVDYEKNTK	
2733.36	2732.35	2732.40	-0.05	305	-	329	1	KTFSHELSDFGLESTTGEIPVVAIR	
2820.22	2819.21	2819.37	-0.16	235	-	258	1	FIQESIFGLCPHHTEDNKDLIQGK	

FIGURE A.3 Identification of peptides recorded on mass spectra.

Matched peptides shown in Bold Red

```
  1 MRFSCLALLP GVALLLASAL LASASDVLEL TDENFESRVS DTGSAGLMLV
 51 EFFAPWCGHC KRLAPEYEAA ATRLKGIVPL AKVDCTANTN TCNKYGVSGY
101 PTLKIFRDGE EAGAYDGPRT ADGIVSHLKK QAGPASVPLR TEDEFKKFIS
151 DKDASVVGFF RDLFSDGHSE FLKAASNLRD NYRFAHTNVE SLVKEYDDNG
201 EGITIFRPLH LANKFEDKIV AYTEKKMTSG KIKKFIQESI FGLCPHMTED
251 NKDLIQGKDL LTAYYDVDYE KNTKGSNYWR NRVMMVAKTF LDAGHKLNFA
301 VASRKTFSHE LSDFGLESTT GEIPVVAIRT AKGEKFVMQE EFSRDGKALE
351 RFLQEYFDGN LKRYLKSEPI PETNEGPVKV VVAESFDDIV NAEDKDVLIE
401 FYAPWCGHCK NLEPKYKELG EKLSKDPNIV IAKMDATAND VPSPYEVKGF
451 PTIYFSPANK KLTPKKYEGG RELNDFISYL QREATNPPII QEEKPKKKKK
501 AQEDL
```

FIGURE A.4   Sequence of protein identified; the red color denotes peptide fragments detected along the mass spectrum. (*See insert for color representation of figure.*)

to define the mass range within which the search should be performed (Figure A.5). Similar to the MASCOT search, PeptIdent identified the protein as disulfide isomerase A3 precursor PDA3 (Figure A.6).

The results obtained here are slightly different from those obtained using MASCOT:

- 39 peptides from PDA3 were identified (34 in MASCOT). The difference arises due to the different search algorithms. For example, peptides having m/z values of: 1284.1189, 1397.6177, 1652.6891, 1653.6762, 1780.9344, 2605.2548, and 2717.7311 revealed by PeptIdent did not occur in MASCOT. The results of the ExPASy database search are shown in Figure A.7.
- The sequence coverage was 81.3% (66% in MASCOT).
- The table in Figure A.7 shows sequences containing SH groups modified by iodoacetamide (although these modifications are taken into account in the MASCOT program, they are not shown in the results table).

The PeptIdent program is also able to match peptides that do not belong to a particular peptide map but are present on mass spectra such as trypsin autolysis products, peptides from human keratins, MALDI matrix adducts, and fragments released from other proteins coexisting in the gel spot. Another reason for disagreement between peptide masses generated theoretically and sample could be, for example, protein modifications, methionine oxidation, or acrylamide adducts. In this type of analysis, the options available on the ExPASy server—FindPept, FindMod, GlycoMod—might be very helpful. A detailed description may be found at www.expasy.org (Proteomics and Sequence Analysis Tools), as shown in Figure A.8.

Preparation of the Report: Following the example shown above, identify the protein selected on the basis of its peptide map mass spectra using the program available

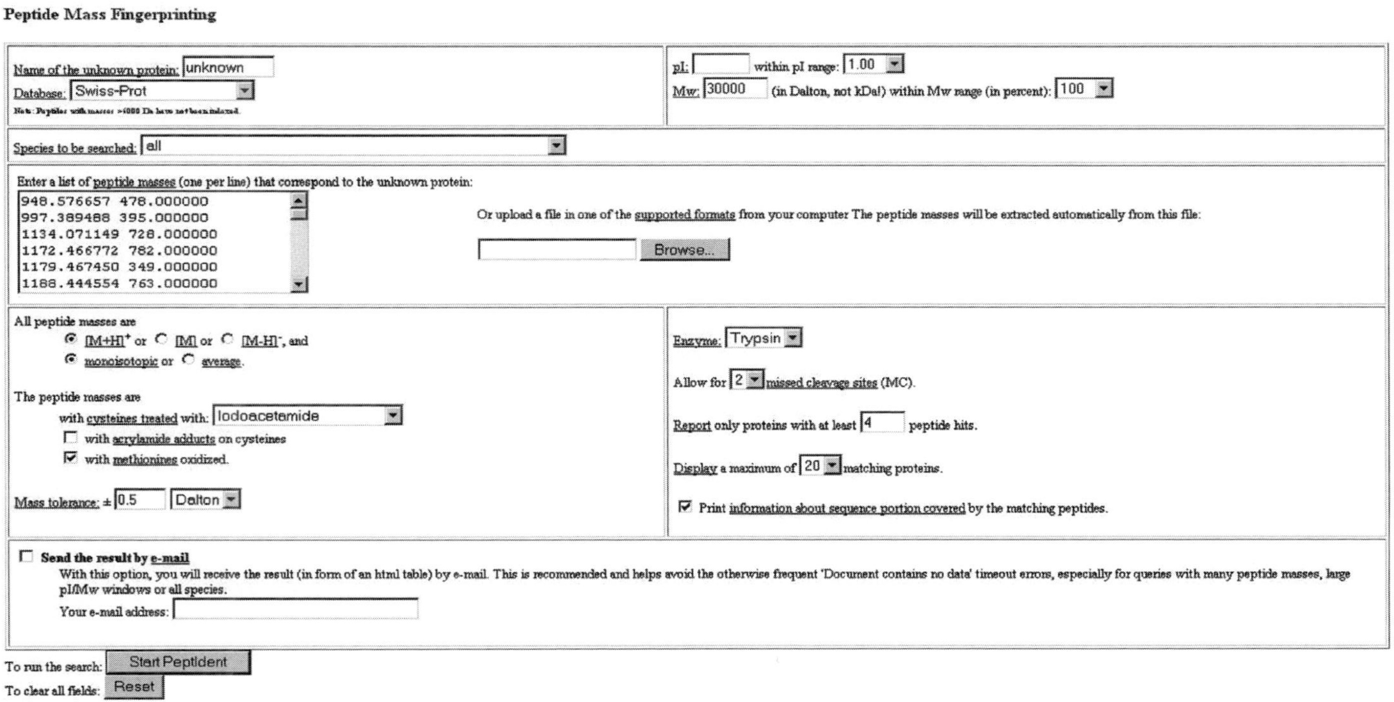

Peptide Mass Fingerprinting

Name of the unknown protein: [unknown]

Database: [Swiss-Prot]

Note: Peptides with masses >499 Da have not been indexed.

pI: [ ] within pI range: [1.00 ▼]

Mw: [30000] (in Dalton, not kDal) within Mw range (in percent): [100 ▼]

Species to be searched: [all ▼]

Enter a list of peptide masses (one per line) that correspond to the unknown protein:

```
948.576657 478.000000
997.389488 395.000000
1134.071149 728.000000
1172.466772 782.000000
1179.467450 349.000000
1168.444554 763.000000
```

Or upload a file in one of the supported formats from your computer. The peptide masses will be extracted automatically from this file:

[ ] [Browse...]

All peptide masses are
⦿ [M+H]+ or ○ [M] or ○ [M-H]+, and
⦿ monoisotopic or ○ average.

The peptide masses are

with cysteines treated with: [Iodoacetamide ▼]
☐ with acrylamide adducts on cysteines
☑ with methionines oxidized.

Mass tolerance: ± [0.5] [Dalton ▼]

Enzyme: [Trypsin ▼]

Allow for [2 ▼] missed cleavage sites (MC).

Report only proteins with at least [4] peptide hits.

Display a maximum of [20 ▼] matching proteins.

☑ Print information about sequence portion covered by the matching peptides.

☐ Send the result by e-mail
With this option, you will receive the result (in form of an html table) by e-mail. This is recommended and helps avoid the otherwise frequent 'Document contains no data' timeout errors, especially for queries with many peptide masses, large pI/Mw windows or all species.

Your e-mail address: [ ]

To run the search: [Start PeptIdent]
To clear all fields: [Reset]

FIGURE A.5 PeptIdent form (see the text for details).

285

Score	#peptide matches	AC	ID	Description	pI	Mw
0.55	39	P11598	PDA3_RAT_1	CHAIN 1: Protein disulfide isomerase A3. - Rattus norvegicus (Rat).	5.78	54239.39
0.30	21	P27773	PDA3_MOUSE_1	CHAIN 1: Protein disulfide isomerase A3. - Mus musculus (Mouse).	5.78	54210.35
0.27	19	Q04638	YMG8_YEAST	Hypothetical 54.1 kDa protein in DAK1-ORC1 intergenic region. - Saccharomyces cerevisiae (Baker's yeast).	5.41	54095.83
0.25	18	P00742	FA10_HUMAN_1	CHAIN 1: Coagulation factor X. - Homo sapiens (Human).	5.38	50336.41
0.25	18	Q06442	WN5A_AMBME_1	CHAIN 1: Wnt-5a protein. - Ambystoma mexicanum (Axolotl).	8.66	38065.24
0.25	18	P41221	WN5A_HUMAN_1	CHAIN 1: Wnt-5a protein. - Homo sapiens (Human).	8.66	38560.86
0.25	18	Q9QXQ7	WN5A_RAT_1	CHAIN 1: Wnt-5a protein. - Rattus norvegicus (Rat).	8.78	38518.87
0.25	18	Q59061	YO67_METJA	Hypothetical protein MJ1667. - Methanococcus jannaschii	9.19	48803.64

FIGURE A.6   List of peptides identified by the system together with their scores. Note that the score value differs from that generated by MASCOT. Refer to Chapter 7.5 for details.

on the Internet. Specify which of the peptides comes from trypsin autolysis, keratins, and nonspecific cleavage. Which peptides contain oxidized methionine or polyacrylamide adducts?

4. PROTEOMICS APPLICATIONS OF LC-MS AND LC-MS/MS

Objectives: To become familiar with the basic operation of a capillary LC system linked to an MS detector. Additionally, data processing and interpretation are emphasized.

Topics to Prepare: Basics of liquid chromatography, types of column packings, modes of chromatography, detectors used in HPLC.

Reagents and Equipment: Bruker Esquire 3000 mass spectrometer with nano-ESI source, Ulimate micro chromatograph and Switchos microcolumn switching module, PepMap reversed-phase capillary column (all from LC Packings/Dionex).

Solvents: Water, acetonitrile, trifluoroacetic acid.

Other Reagents: Trypsin, iodoacetamide, dithiotreitol (DTT).

Procedure

1. Familiarize yourself with a capillary LC-MS system and discuss the setup with your instructor.
2. Be aware of the high voltage applied to the nano-ESI interface.
3. Inject the lysate received.
4. Set up the correct fragmentation parameters.
5. Start gradient separation and data acquisition.
6. Analyze the data recorded using the manufacturer-supplied software.
7. Perform the automatic protein identification procedure using MASCOT software from Matrix Science and interpret the data obtained.

Score: 0.55, 39 matching peptides: P11598 (PDA3_RAT) pI: 5.78, Mw: 54239.39
CHAIN 1: Protein disulfide isomerase A3 - Rattus norvegicus (Rat).

| GlycoMod | FindMod | FindPept | PeptideMass | BioGraph |

user mass	matching mass	Δmass (Dalton)	#MC	modification	position	peptide
948.5767	948.5118	-0.0648	1		281-288	NRVMMVAK
997.3895	997.5101	0.1206	0		153-161	DASVVGFFR
1172.4668	1172.5405	0.0737	0		336-344	FVMQEEFSR
1179.4675	1179.5865	0.119	1		174-183	AASNLRDNYR
1188.4446	1188.5354	0.0908	0	MSO: 338	336-344	FVMQEEFSR
1191.6078	1191.6004	-0.0073	0		63-73	LAPEYEAAATR
1194.6358	1194.6729	0.0371	2		416-425	YKELGEKLSK
1236.4681	1236.5127	0.0446	0		108-119	DGEEAGAYDGPR
1244.5771	1244.6634	0.0863	0		184-194	FAHTNVESLVK
1284.1189	1283.5355	-0.5833	0		83-94	VDCTANTNTCNK
1341.604	1341.6838	0.0798	0		449-460	GFPTIYFSPANK
1342.621	1342.6889	0.0679	1		215-225	FEDKIVAYTEK
1373.6019	1373.6736	0.0717	0		352-362	FLQEYFDGNLK
1394.597	1394.6587	0.0617	0		162-173	DLFSDGHSEFLK
1395.6505	1395.6876	0.0371	2		272-282	NTKGSNYWRNR
1396.5647	1396.6954	0.1307	0		367-379	SEPIPETNEGPVK
1397.6177	1397.5784	-0.0392	0	2xCys_CAM	83-94	VDCTANTNTCNK
1397.6177	1397.706	0.0883	0		472-482	ELNDFISYLQR
1469.6524	1469.7787	0.1263	1		449-461	GFPTIYFSPANKK
1488.6838	1488.6787	-0.005	1	MSO: 338	336-347	FVMQEEFSRDGK
1529.6964	1529.7747	0.0783	1		352-363	FLQEYFDGNLKR
1547.2348	1547.8073	0.5725	1		283-296	VMMVAKTFLDAGHK
1587.7407	1587.8165	0.0758	1		148-161	FISDKDASVVGFFR
1588.742	1588.2805	0.1385	2		62-75	RLAPEYEAAATRLK
1652.6891	1652.7472	0.0581	0	MSO: 434	434-448	MDATANDVPSPYEVK
1652.6891	1652.7663	0.0772	1		105-119	IFRDGEEAGAYDGPR
1653.6762	1653.7602	0.084	0		25-38	SDVLELTDENFESR
1715.8715	1715.9115	0.04	2		147-161	KFISDKDASVVGFFR
1721.7834	1721.9432	0.1578	1		483-497	EATNPPIIQEEKPKK
1744.9069	1744.8864	-0.0204	1		131-146	QAGPASVPLRTEDEF K
1746.8636	1746.9286	0.065	1		289-304	TFLDAGHKLNFAVAS R
1749.9178	1749.8541	-0.0636	0		380-395	VVVAESFDDIVNAED K
1780.9344	1780.8185	-0.1158	0		396-410	DVLIEFYAPWCGHCK
1792.7308	1792.8977	0.1669	1		180-194	DNYRFAHTNVESLVK
1801.0591	1800.9378	-0.1212	1		364-379	YLKSEPIPETNEGPV K
1872.9192	1872.9814	0.0622	2		130-146	KQAGPASVPLRTEDE FK
1872.9192	1872.9814	0.0622	2		131-147	QAGPASVPLRTEDEF KK
2302.0771	2302.1462	0.0691	0		195-214	EYDDNGBOITIFRPL HLANK
2605.2548	2605.3032	0.0484	2		253-274	DLIQGKDLLTAYYDY DYEKNTK
2605.2548	2605.3144	0.0596	0		306-329	TFSHELSDFGLESTT GEIPVVAIR
2717.7311	2718.3158	0.5847	2		95-119	YGVSGYPTLKIFRDG EEAGAYDGPR
2733.359	2733.4093	0.0503	1		305-329	KTFSHELSDFGLEST TGEIPVVAIR
2820.2194	2820.3695	0.1501	1	1xCys_CAM	235-258	FIQESIFGLCPHMTE DNKDLIQGK

ΔMw: 24239.4 Da (80.8%)
81.3% of sequence covered:

```
          1        11        21        31        41        51
          |        |         |         |         |         |
  1                           SDVLEL TDENFESRvs dtgsaglmlv effapwcghc   60
 61 kRLAPEYEAA ATRLKgivpl akVDCTANTN TCNKYGVSGY PTLKiFRDGE EAGAYDGPRt  120
121 adgivshlkK QAGPASVPLR TEDEFEKFIS DKDASVVGFF RDLFSDGHSE FLKAASNLRD  180
181 NYRFAHTNVE SLVKEYDDNG BGITIFRPLH LANKFEDKIV AYTEKnatsg kikkFIQESI  240
241 FGLCPHMTED NKDLIQGKDL LTAYTDVDYE KNTHGSNYWR NRGMNVAKTF LDAGHKLNFA  300
301 VASRKTFSHE LSDFGLESTT GEIPVVAIRt akgekFVMQE EFSRDGKale rFLQEYFDGN  360
361 LKRYLKSEPI PETNEGPVKV VVAESFDDIV NAEDMDVLIE FYAPWCGHCK nlepkYKELG  420
421 EKLSKdpniv iakMDATAND VPSPYEVKGF PTIYFSPANK Kltpkkyegg rELNDFISYL  480
481 QREATNPPII QEEKPKGkkk aqedl
```

FIGURE A.7   Results of the search in the ExPASy database using the PeptIdent program. The sequence of the identified protein is given below. Note the accuracy of the calculations and measurements (four decimal digits), which cannot be obtained during low-resolution analysis. (*See insert for color representation of figure.*)

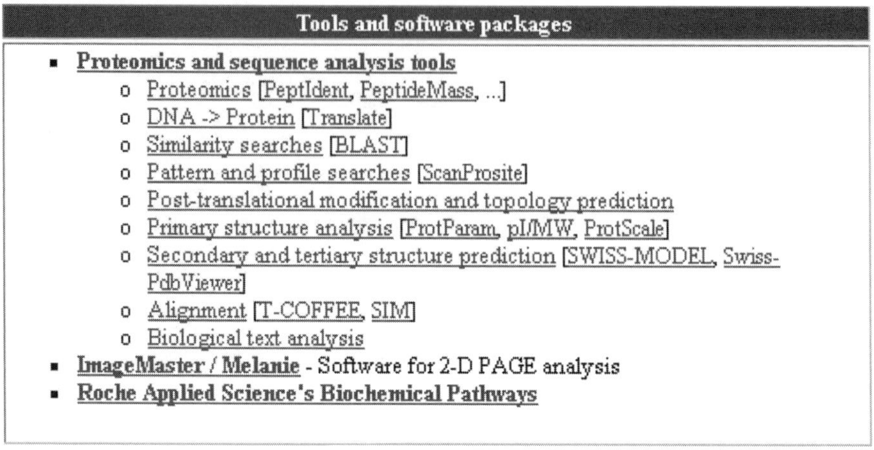

Tools and software packages

■ **Proteomics and sequence analysis tools**
   ○ Proteomics [PeptIdent, PeptideMass, ...]
   ○ DNA -> Protein [Translate]
   ○ Similarity searches [BLAST]
   ○ Pattern and profile searches [ScanProsite]
   ○ Post-translational modification and topology prediction
   ○ Primary structure analysis [ProtParam, pI/MW, ProtScale]
   ○ Secondary and tertiary structure prediction [SWISS-MODEL, Swiss-PdbViewer]
   ○ Alignment [T-COFFEE, SIM]
   ○ Biological text analysis
■ **ImageMaster / Melanie** - Software for 2-D PAGE analysis
■ **Roche Applied Science's Biochemical Pathways**

FIGURE A.8   ExPASy server link page.

Preparation of the Report

1. Compare the UV-Vis and MS detectors linked to the LC system and specify their pros and cons.
2. Discuss the biological function of the protein identified. Use other databases (e.g., Ovid, PubMed) for reference retrieval.
3. Discuss sources of possible errors that could influence your results.

5. ENZYME KINETICS

Objectives: To learn how neuropeptides are metabolized in the nervous tissue, and to acquire skills in monitoring the formation of shorter sequences and estimate the basic kinetic parameters of the reaction.

Topics to Prepare

- Enzymes, basic mechanisms of action, classification, kinetics, Michaelis constant, Michaelis–Menten equation, Lineweaver–Burk plot, allosteric enzymes, determination of kinetic parameters
- Active site and catalytic mechanism of serine proteinases
- Inhibitors (families of proteinase inhibitors, reversible and irreversible inhibitors, competitive and noncompetitive inhibitors)

Reagents and Equipment: Mass spectrometer with an ESI source, and a manual injection valve.

Reagents: rat brain homogenate, nociceptin/orphanin FQ (1 mg/mL, water solution), amastatin (0.6 mM, water solution), Tris-HCl (20 mM, pH 8.4, water solution), 30% methanol in water, supplemented with 0.1% formic acid.

Procedure

1. Thaw the rat brain homogenate received. This has been prepared by homogenization of the tissue in 10 mM Tris-HCl buffer, pH 7.8, in a glass–glass homogenizer centrifuged to remove cell debris. The tissue-to-buffer ratio is about 1 : 3 w/w.
2. Dilute this solution 20 times (v/v) with Tris-HCl buffer, pH 8.4.
3. Prepare the following water solutions of neuropeptide nociceptin/orphanin FQ: 1 mg/mL, 0.7 mg/mL, 0.5 mg/mL, 0.3 mg/mL, and 0.1 mg/mL in 150 to 200-μL volumes.
4. Dissolve 10 times the amastatin stock solution (with water or a buffer).
5. Label 0.5-mL Eppendorf tubes for peptide solutions; also prepare an additional empty tube labeled "kinetics."
6. To the labeled tubes add 20 μL of amastatin and 20 μL of diluted rat brain homogenate.
7. Incubate the tubes for 20 minutes at 37°C (in a water bath or thermoblock). **Caution:** The tube cap may not hold, due to the slight overpressure caused by the increased temperature.
8. Add 20 μL of the appropriate substrate solutions. To the additional tube ("kinetics") add 20 μL of 0.5 mg/mL substrate solution.
9. Incubate the labeled tubes for 60 minutes at 37°C. Incubate the additional tube ("kinetics") at the same temperature, but remove 5-μL aliquots every 10 minutes (including $t = 0$ minutes). Transfer the samples withdrawn into new tubes and add 10 μL of 30% ethanol with 0.1% formic acid. Freeze all tubes.
10. Analyze all the solutions in the mass spectrometer with electrospray ionization and save the data recorded.

Preparation of the Report

1. Discuss the procedure and explain all the major steps.
2. What was the rationale for incubating samples for 60 minutes and withdrawing aliquots after time intervals?
3. What types of data will be used for quantitative and qualitative calculations?
4. Based on the data obtained from the time-dependence experiment (tube labeled "kinetics"), determine all products formed during incubation and plot the curve describing how fast they occurred (dependence between the quantity of product and the time of incubation, in arbitrary units).

5. Based on the data obtained from samples containing increasing substrate concentrations, plot the Lineweaver–Burk curve and calculate K_M and V_{max}.

6. INTERNET DATABASES

Objectives: To reveal the sequence and function of the protein selected using bioinformatic tools available on the Internet.

Topics to Prepare: Databases applied in proteomics and their properties, content, and applications, basic bioinformatics tools and their applications.

Procedure

SwissProt Database

1. Find the human κ-opioid receptor in the SwissProt database (http://www. expasy.org). Save its entry name and find its primary accession number. Locate basic information about this protein: protein family, number of amino acids, molecular mass, and the name of the gene that encodes this protein.
2. In the *Features* section, locate basic information about domains of which this protein is composed. How many transmembrane domains can you find? Which protein fragments are located in cytoplasm, and which stick out of the cell? Are there any known posttranslational modifications for the κ receptor? In which domains are they located?

Bioinformatic Tools

1. Find proteins with a sequence similar to the sequence of κ receptor using the BLAST program available from the SwissProt database. Find differences between the human, rat, and mouse κ receptors. How many amino acids in the sequences are identical, and how many are similar? In which parts of the sequences are the differences biggest?
2. Compare the human κ receptor with the human μ and δ receptors. Are there any particularly similar or different fragments? Which of them are these?
3. Find proteins similar to the κ receptor in the crystallographic structure database (PDB) using the BLAST program available at the EBI Website (http://www.ebi.ac.uk./blast). Export the receptor sequence edited in the FASTA format. Is there any close sequence similarity between the protein(s) you found and the κ receptor that are sufficient for their structural similarity?

Biological Information Search

1. Search for the human receptor κ entry in the Ensembl genomic database (http://www.ensembl.org). You can also use links from the SwissProt database

entry (cross-references). How many exons and introns is the κ receptor gene composed of? On which chromosome is the gene located?

2. Find the κ receptor description using the Interpro protein domain database (http://www.ebi.ac.uk/interpro). Which chemical process is this protein involved in? In which parts of the central nervous system does it show the highest abundance? To which protein family and superfamily does it belong? Does the description of the receptor confirm conclusions drawn from the sequences compared by the BLAST method?

Preparation of the Report: The report should contain results obtained during the experiment and answers to questions stated in the text.

APPENDIX B

DATA TABLES

TABLE B.1 Selected Posttranslational Modifications

Modification	Modification Place	Mass Difference	
		Monoisotopic	Average
Acetylation	N-terminal, Lys	42.010	42.037
Palmitoylation	Cys, Lys, Ser, Thr	238.229	238.408
Carbamidomethyl	C-terminal, Asp, Glu	57.0340	57.072
Carboxymethyl	C-terminal, Asp, Glu	58.005	58.037
Farnesylation	Cys	204.188	204.351
Myristoylation	Gly (N-terminal), Lys	210.198	210.355
Biotinylation	N-terminal, Lys	226.078	226.295
Deamidation	Asn, Gln	0.984	0.985
Phosphorylation	Ser, Thr, Tyr	79.966	79.980
Formylation	N-terminal, Lys	27.995	28.010
Propionamide	C-terminal, Asp, Glu	71.037	71.079
Pyro-glu	N-terminal, Gln	−17.026	−17.030
Pyro-glu	N-terminal, Glu	−18.010	−18.015
S-Pyridylethyl	C-terminal, Asp, Glu	105.057	105.139
Glycosylation			
N- glycosylation	Asn-X-Ser(Thr)		>800
N-acetylglucosamine	Asn	203.079	203.192
O-glycosylation	Ser, Thr		>800
Hydroxylation	Asp, Lys, Asn, Pro	15.995	15.999
Methylation	N-terminal, Cys, His, Lys, Asn, Gln, Arg	14.016	14.027
Ubiquitinylation	Lys	114.042	114.103
Oxidation	His, Met, Trp	15.995	15.999
Sulfonation	Met	31.990	31.999

Proteomics: Introduction to Methods and Applications, Edited by Agnieszka Kraj and Jerzy Silberring
Copyright © 2008 John Wiley & Sons, Inc.

TABLE B.2 Masses and Structures of Commonly Occurring and Some Less Common Amino Acid Residues

Name	Code	Residue	Monoisotopic[a]	Average[b]
			Mass (Da)	
Alanine	Ala (A)	CH_3 $\|$ —NH—CH—CO—	71.03711	71.0779
2-Aminobutyric acid	2-Aba	CH_2—CH_3 $\|$ —NH—CH—CO—	85.05276	85.1045
Aminoethyl-cysteine	AECys	CH_2—S—$(CH_2)_2$—NH_2 $\|$ —NH—CH—CO—	146.05138	146.2107
2-Aminoisobu-tyric acid	Aib	CH_3 $\|$ —NH—C—CO— $\|$ CH_3	85.05276	85.1045
Arginine	Arg (R)	H_2C—$(CH_2)_2$—NH—C—NH_2 $\|$ $\|\|$ —NH—CH—CO— NH	156.10111	156.1857
Asparagine	Asn (N)	CH_2—$CONH_2$ $\|$ —NH—CH—CO—	114.04293	114.1026
Aspartic acid	Asp (D)	CH_2—COOH $\|$ —NH—CH—CO—	115.02694	115.0874
Carboxymethyl-cysteine	Cmc	CH_2—S—CH_2—COOH $\|$ —NH—CH—CO—	161.01466	161.1790
4-Carboxyglu-tamic acid	Gla	COOH $\|$ CH_2—CH—COOH $\|$ —NH—CH—CO—	173.03242	173.1235
Cysteic acid	Cys(O_3H)	CH_2—SO_3H $\|$ —NH—CH—CO—	150.99393	151.1411
Cysteine	Cys (C)	CH_2—SH $\|$ —NH—CH—CO—	103.00918	103.1429

TABLE B.2 Masses and Structures of Commonly Occurring and Some Less Common Amino Acid Residues (*Continued*)

Name	Code	Residue	Mass (Da) Monoisotopic[a]	Average[b]
Dehydroalanine	Dha	CH_2 $\parallel$ —NH—C—CO—	69.02146	69.0620
2-Dehydro-2-aminobutyric acid	Dhb	$CH—CH_3$ $\parallel$ —NH—C—CO—	83.03711	83.0886
Glutamic acid	Glu (E)	$CH_2—CH_2—COOH$ | —NH—CH—CO—	129.04259	129.1140
Glutamine	Gln (Q)	$CH_2—CH_2—CONH_2$ | —NH—CH—CO—	128.05858	128.1292
Glycine	Gly (G)	—NH—CH_2—CO—	57.02146	57.0513
Histidine	His (H)	H_2C ... —NH—CH—CO—	137.05891	137.1393
Homocysteine	Hcy	$CH_2—CH_2—SH$ | —NH—CH—CO—	117.02483	117.1695
Homoseryne	Hse	$CH_2—CH_2—OH$ | —NH—CH—CO—	101.04768	101.1039
5-Hydroxylysine	Hyl	OH | $H_2C—CH_2—CH—CH_2—NH_2$ —NH—CH—CO—	144.08988	144.1717
4-Hydroxyproline	Hyp	OH ... —N—HC—CO—	113.04768	113.1146

(*Continued*)

TABLE B.2 Masses and Structures of Commonly Occurring and Some Less Common Amino Acid Residues (*Continued*)

Name	Code	Residue	Mass (Da)	
			Monoisotopic[a]	Average[b]
Isoleucine	Ile (I)	H₃C—CH—CH₂—CH₃ / —NH—CH—CO—	113.08406	113.1576
Isovaline	Iva	H₂C—CH₃ / —NH—C—CO— / CH₃	99.06841	99.1311
Leucine	Leu (L)	H₂C—CH(CH₃)₂ / —NH—CH—CO—	113.08406	113.1576
Lysine	Lys (K)	H₂C—(CH₂)₃—NH₂ / —NH—CH—CO—	128.09496	128.1723
Methionine	Met (M)	H₂C—CH₂—S—CH₃ / —NH—CH—CO—	131.04048	131.1961
Norleucine	Nle	CH₂—(CH₂)₂—CH₃ / —NH—CH—CO—	113.08406	113.1576
Norvaline	Nva	H₂C—CH₂—CH₃ / —NH—CH—CO—	99.06841	99.1311
Ornithine	Orn	H₂C—(CH₂)₂—NH₂ / —NH—CH—CO—	114.07931	114.1457
Phenylalanine	Phe (F)	H₂C— (phenyl ring) / —NH—CH—CO—	147.06841	147.1739
2-Piperidine-carboxylic acid	Pip	(piperidine ring) / —N—CH—CO—	111.06841	111.1418
Proline	Pro (P)	(pyrrolidine ring) / —N—HC—CO—	97.05276	97.1152
Pyroglutamic acid	pGlu	O= (ring) / —N—CH—CO—	111.03203	111.0987

TABLE B.2 Masses and Structures of Commonly Occurring and Some Less Common Amino Acid Residues (*Continued*)

Name	Code	Residue	Monoisotopic[a]	Average[b]
			Mass (Da)	
Sarcosine	Sar	CH$_3$ | —N—CH$_2$—CO—	71.03711	71.0779
Serine	Ser (S)	CH$_2$—OH | —NH—CH—CO—	87.03203	87.0773
Threonine	Thr (T)	HO—CH—CH$_3$.| —NH—CH—CO—	101.04768	101.1039
Tryptophan	Trp (W)	H$_2$C— (indole ring, NH) —NH—CH—CO—	186.07931	186.2099
Tyrosine	Tyr (Y)	H$_2$C— (benzene ring) —OH —NH—CH—CO—	163.06333	163.1733
Valine	Val (V)	H$_3$C—CH—CH$_3$ | —NH—CH—CO—	99.06841	99.1311

[a] Monoisotopic masses were calculated on the basis of the atomic masses of the most abundant isotopes of elements: C, 12.000000 Da; H, 1.007825 Da; N, 14.003074 Da; O, 15.994915 Da; S, 31.972070 Da.
[b] Average masses were calculated on the basis of the weighted-average atomic masses of elements (taking account of the content of each isotope): C, 12.0107 Da; H, 1.00794 Da; N, 14.0067 Da; O, 15.9994 Da; S, 32.065 Da.

TABLE B.3 Pressure Units of Measurement

Unit	Abbreviation	Comments
pascal	Pa	SI unit
bar	bar	10^5 Pa
millimeter of mercury	mmHg	133,322 Pa
torr	torr	1 torr = 1 mmHg (at 0°C)
atmosphere	atm	101,325 Pa = 760 mmHg
technical atmosphere	at	9.80665×10^4 Pa
pound-force per square inch	psi	6895 Pa

TABLE B.4   Products of Trypsin Autolysis

Monoisotopic Mass (Da)	Fragment[a]	MC[b]
5191.4975	1–49	1
4551.1318	1–43	0
4417.1980	50–89	1
3686.8040	100–136	1
3670.5916	150–186	1
3338.6467	92–125	1
3195.4795	140–170	1
2910.3178	171–200	1
2803.4220	44–69	1
2552.2483	100–125	0
2514.3384	70–91	1
2321.2070	187–208	1
2273.1594	70–89	0
2193.9943	150–170	0
2163.0564	50–69	0
1999.0469	201–217	1
1725.8628	209–223	1
1497.7617	126–139	1
1495.6151	171–186	0
1433.7205	187–200	0
1364.6912	137–149	1
1153.5735	126–136	0
1111.5605	209–217	0
1046.5953	90–99	1
1020.5030	140–149	0
906.5043	201–208	0
805.4162	92–99	0

Source: Data from Swiss-Prot database. PeptideMass tool was used for bovine
trypsin (P00760). Only peptides of masses greater than 750 Da are shown.
[a]Numbering refers to the active form of protein.
[b]MC, missed cleavages.

APPENDIX C

INTERNET RESOURCES

Filip Sucharski, Justyna Jarzebinska, and Hana Raoof

LITERATURE DATABASES

http://www.ncbi.nlm.nih.gov/literature

Databases of the National Center for Biotechnology Information, including PubMed (MEDLINE biomedical literature), PubMed Central (free access to the full text of life science journals), Books (collection of online biomedical books), OMIM (Online Mendelian Inheritance in Man, catalog of inherited diseases), OMIA (Online Mendelian Inheritance in Animals, catalog of genes and genetic diseases in animals), and others.

http://gateway.nlm.nih.gov

Web-based system that lets users search simultaneously in databases provided by the National Library of Medicine. Among the others: PubMed, OMIM (mentioned above), TOXLINE Subset (toxicology references), and MedlinePlus Medical Encyclopedia.

http://books.google.com/

Beta version of Google Book Search engine. Allows users to browse the full text of books to get to know where to buy or borrow them. Some books are available in full view (those out of copyright or which the author or publisher has made fully viewable); the rest can be seen in limited preview, or only information about book is available.

Proteomics: Introduction to Methods and Applications, Edited by Agnieszka Kraj and Jerzy Silberring
Copyright © 2008 John Wiley & Sons, Inc.

http://scholar.google.com

Google Scholar enables searching texts of scientific journals, articles, theses, abstracts, books, and the like.

http://www.scirus.com/

Science-specific search engine, which focuses on Web pages containing scientific content. It is possible to define the type of information that is searched (abstract, article, book, conference, thesis, disssertation), the source (journal, Web), the subject area, or the file format.

http://www.scopus.com

Abstract and citation database of research literature and high-quality Web resources. Covers open-access journals, conference proceedings, and books. Contains titles from *Nature*, Biomed Central, Springer-Verlag, and many others, even non-English titles.

http://www.sciencedirect.com/

Offers online medical and technical information from over 2000 Elsevier journals and about 4000 books from physical, life, health sciences, engineering, or social sciences and humanities subject areas.

http://www.biomedcentral.com/

Free access to nearly 200 journals containing peer-reviewed biomedical articles. Previous registration needed.

http://www.springerlink.com/

Articles on biomedical and life sciences, medicine, chemistry, and material science and other subject matter from journals and books published by the Springer Science + Business Media group. Free access only to abstracts; full versions are available for licensed users.

http://www.ingentaconnect.com/

Academic and professional research articles in various subject areas: from medicine, chemistry, biology, life sciences through mathematics, and economics to psychology, philosophy, and the arts.

http://www.pnas.org/

Site containing articles published by the *Proceedings of the National Academy of Sciences of the United States of America*. Some articles are available free; all are available only to licensed users.

http://www.interscience.wiley.com

Archive of the Wiley Interscience publishing house dating back to 1799, containing over 1.5 million articles from more than 2500 journals. For licensed users only.

http://www.nature.com/

Site that offers searching by key words from the article base of the Nature Publishing Group. Some of the papers are freely available.

USEFUL WEB SITES

http://www.abrf.org
The Association of Biomolecular Resource Facilities. Contains an interesting discussion forum covering a wide range of biotechnological problems. It also comprises Delta Mass, a database of posttranslational modifications.

http://www.chem.uky.edu/research/lynn/proteomics1.pdf
Molecular Biologist's Guide to Proteomics. An article describing the principles of proteomics and typical analytical methods for proteome analysis.

http://www.expasy.org
ExPASy (Expert Protein Analysis System) proteomics server. Dedicated to the analysis of protein structures and sequences as well as two-dimensional PAGE. It contains a wide range of protein databases (e.g., SwissProt) and useful proteomics tools such as PeptIdent or software for two-dimensional PAGE gel analysis.

http://www.matrixscience.com
Matrix Science comprises a powerful search engine coupled to the largest protein databases. It allows protein identification on the grounds of mass spectra.

http://www.pubmed.com
Most popular database for biomedical articles and abstracts.

http://prospector.ucsf.edu/
Protein Prospector provides a number of useful bioinformatical tools (e.g., MS-Digest, MS-Tag, MS-Fit) which are important in dealing with proteomics.

http://www.spectroscopynow.com/proteomics/
Subfolder in a large portal dedicated to spectroscopy. The site contains a number of interesting articles concerning proteomics, educational facts, and news; also available in the RSS mode.

http://www.separationsnow.com
Site dedicated to analytical methods for separation. Contains information about electrophoresis, chromatography, and proteomics.

http://www.unimod.org
Site containing detailed information concerning protein modifications, particularly for mass spectrometry.

http://prowl.rockefeller.edu/
PROWL contains a number of bioinformatic tools for protein handling (e.g., Profound, Peptidemap).

http://www.ebi.ac.uk/
Site of the European Bioinformatics Institute, which provides databases, tools, and a lot of proteomics and genomics tutorials, particularly at http://www.ebi.ac.uk/2can/tutorials/index.html.

http://www.eupa.org

European Proteomics Association Provides detailed information about conferences, meetings, trainings, and job offers in the field of proteomics.

http://www.hupo.org

Human Proteome Organization, which is an international scientific organization promoting proteomics through international cooperation and collaborations. The site provides information about congresses, meetings, trainings, and common projects that are realized in laboratory networks.

http://www.ocms.ox.ac.uk/links/Biochemistry.html

Huge set of links from the field of biochemistry, focused primarily on proteomics and genomics.

http://www.hprd.org/

Human Protein Reference Database. The site contains the BLAST tool and is useful in phosphoproteome analysis. The PhosphoMotif Finder makes it possible to check if a protein contains any phosphorylation site described previously in the literature.

http://ncrr.pnl.gov/software/

Site comprising proteomics-dedicated software, other data files from MS analysis, and some protocols.

http://ionsource.com/

Site containing a wide range of data sets, from a peptide mass calculator to an HPLC clip art gallery. There are some noteworthy tutorials.

http://www.cbs.dtu.dk/services/NetPhos/

Server producing neural network predictions for Ser, Thr, and Tyr phosphorylation sites in eukaryotic cells.

http://www.phosida.de

Phosphorylation site database.

http://peptide.ucsd.edu/MassSpec

Tool for mass spectrum interpretation of de novo sequenced peptides.

http://peptide.ucsd.edu/MassSpec

Open-access online journal covering proteomic research.

BIOINFORMATICS DATABASES

Nucleotide Sequence Databases

http://www.ebi.ac.uk/embl

European Bioinformatics Institute's database.

http://www.ncbi.nlm.nih.gov/Genbank/

National Center of Biotechnological Information's database.

http://www.ddbj.nig.ac.jp/

DNA Data Bank of Japan.

Protein Databases

http://www.expasy.org/sprot

SwissProt; database that provides manually verified data in particular protein sequences.

http://pir.georgetown.edu/

Protein Information Resource (PIR); first international Protein Sequence Database.

http://www.expasy.uniprot.org

Universal Protein Resource (UniProt); combines SwissProt information, TrEMBL, and PIR.

ftp://ftp.ncbi.nih.gov/repository/MSDB

Database directed to MS applications. A guidebook for MSDB users can be found at http://www.matrixscience.com/help/seq_db_setup_msdb.html.

http://www.rcsb.org/pdb

International repository of three-dimensional structures of proteins, nucleic acids, viruses.

http://www.expasy.org/prosite

Prosite; database of protein domains, families, and functional sites.

http://www.ebi.ac.uk/interpro

InterPro; database of protein families, domains, and functional sites; makes possible structure and function prediction on the basis of known protein fragments.

http://merops.sanger.ac.uk

Merops; database of peptidases and their proteinaceous inhibitors.

Others

http://www.matrixscience.com/

MASCOT; interface for database search (e.g., SwissProt, MSDB); uses the MOWSE probability-based scoring algorithm.

http://prospector.ucsf.edu

ProteinProspector; database of search tools.

http://expasy.org/tools/

ExPASy proteomics; tools and online programs for protein identification and characterization, and structure pattern prediction.

http://www.ncbi.nih.gov/entrez

Entrez; integrates the NCBI database with the MEDLINE literature database.

http://metlin.scripps.edu

METLIN metabolite database; a repository for mass spectral metabolite data.

http://www.swepep.com/

SwePep; database of endogenous peptides to help with their identification by MS.

http://www.cbs.dtu.dk/services/SignalP

SignalP; site of the Center for Biological Sequence Analisys (CBS); assists in the determination of signal peptides in a given amino acid sequence.

INDEX

Proteomics: Introduction to Methods and Applications, Edited by Agnieszka Kraj and Jerzy Silberring
Copyright © 2008 John Wiley & Sons, Inc.